THE LIVES OF

BUTTERFLIES

THE LIVES OF BUTTERFLIES

A NATURAL HISTORY OF OUR PLANET'S BUTTERFLY LIFE

David G. James & David J. Lohman

PRINCETON UNIVERSITY PRESS
PRINCETON AND OXFORD

Published by Princeton University Press
41 William Street, Princeton, New Jersey 08540
99 Banbury Road, Oxford OX2 6JX
press.princeton.edu

Library of Congress Control Number 2022948544

ISBN 978-0-691-24056-5
Ebook ISBN 978-0-691-24112-8

Typeset in Bembo and Futura

Printed and bound in Latvia
10 9 8 7 6 5 4 3 2 1

British Library Cataloging-in-Publication Data is available

This book was conceived, designed, and produced by
UniPress Books Limited
Publisher: Nigel Browning
Commissioning editor: Kate Shanahan
Project manager: Kate Duffy
Designer & art director: Wayne Blades
Picture researcher: Brendan O'Connell
Illustrator: Sarah Skeate
Maps: Les Hunt

Cover image: (Front) Ondrej Prosicky / Alamy Stock Photo,
(back cover and spine) Musak / iStock

CONTENTS

INTRODUCTION

The Butterfly Effect

Butterflies are a subgroup of moths in the large order of insects called Lepidoptera. Lepidoptera means "scale-winged," a characteristic that separates them from other kinds of insects. The scales give butterflies their color, but are only loosely attached to the wings and gradually wear off during their lifespan. The colors of each butterfly have evolved to provide protection to the species from predators and to enable the sexes to find and recognize each other.

Many people are captivated and intrigued by the beauty of butterflies—the colorful ambassadors of the insect world. However, their appeal goes beyond mere attractiveness. Indeed, for many cultures they symbolize joy, transformation, rebirth, resilience, and hope. In Greek folklore, for example, the human soul was often represented as a butterfly, which is another meaning of the Greek word *psyche*.

However, beyond their beauty and symbolism, we also want to know how butterflies function and survive in the world today—and what their prospects are for the future. The question is sometimes asked: "What good are butterflies?" and "Why should I care about them?" There are many sound scientific and ecological answers to these questions. First and foremost, butterflies are an essential part of most terrestrial ecosystems and an important link in food chains, consuming plants as caterpillars, pollinating flowers as adults, and being consumed by a community of vertebrates and invertebrates. They are also sensitive ecological indicators: if an ecosystem is degraded, butterflies are one of the first groups to suffer, providing an early warning of a developing problem.

This book will further explore the importance and function of butterflies in ecology and nature by taking you into their daily lives.

LEPIDOPTERISTS

The renowned lepidopterist Lincoln Brower summed up the appeal of butterflies: "Butterflies are treasures, like great works of art. Should we not value them as much as the beauty of Picasso's art or the music of Mozart or the Beatles?" Lepidopterists are people who collect and study butterflies. In Victorian times, collectors caught butterflies in nets, killed them, and pinned them to boards in collections. A lot of valuable information on butterfly diversity came from these types of collections,

↑   A Red-spotted Purple (*Limenitis arthemis*) butterfly.

→   A Small Tortoiseshell (*Aglais urticae*) butterfly feeds on the much-maligned but ecologically important stinging nettle.

and some collecting is still crucial for scientific purposes. However, the modern butterfly collector is more likely to be carrying a camera than a net, and their collection comprises thousands of butterfly images rather than actual specimens. Rather than simply collecting butterflies as objects of beauty, we now want to know about their ecology, how they function, and how they fit into ecosystems. We want to know how they are faring in a world humans have shaped and damaged. And we want to know how we can help them.

There are around 19,500 described butterfly species in the world and many more yet to be described, each with a unique lifestyle and strategy for survival. The study of butterfly lives is relatively young, and although we know a lot about some species, there is still much to learn about others. This book will introduce you to some of the celebrated butterfly lives we know about, as well as some of the lesser-known ones.

EDUCATION AND ADVOCACY

Butterfly populations are declining worldwide. Not all species are suffering, but many are. The three major drivers of butterfly decline are habitat loss, pesticide use, and climate change. Most people live busy lives and do not know that butterflies are in trouble, but anyone can help butterflies.

The "extinction of experience," a concept coined by pioneer butterfly conservationist Robert Michael Pyle, describes a child's loss of interaction with nature. This can have negative effects on human health and well-being, and it has also been shown to reduce support for pro-biodiversity policies and programs later in life. Although our love for butterflies appears to be innate,

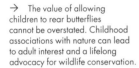 ← Capturing butterflies today is more likely done with a camera than a net.

→ The value of allowing children to rear butterflies cannot be overstated. Childhood associations with nature can lead to adult interest and a lifelong advocacy for wildlife conservation.

the experiences many of us had with them as children appear to play a big role in our appreciation of them later in life. Many older readers will have kept tadpoles and caterpillars when they were young, which often sparked fascination and a lifelong love of nature. Are there as many children today experiencing these wonders of life?

This is why it is so important to allow children to be close to nature and to answer their questions. It is a well-worn cliché, but our children are the future and it is they who will determine the future of butterflies. If a child finds a caterpillar, let them keep it, feed it, and watch it metamorphose. They will remember the experience for the rest of their life, and it will instill in them a love and appreciation for lives smaller than their own.

REGULATION IS NOT THE WAY

Preventing the extinction of experience may well be one of the most important things we can do for the conservation of butterflies. In the past, butterflies classified as endangered were species that were intrinsically rare, with limited distributions and restricted habitats. In contrast, many butterflies that are declining in numbers today are formerly common species that have large ranges and occupy a wide range of habitats. A good example is the iconic Monarch butterfly (*Danaus plexippus*).

The population of the Monarch in North America is estimated to have declined by 80–90 percent during the past two decades, caused by a combination of habitat loss, pesticide use, and a warming climate. The loss of this butterfly is causing

concern, and numerous programs have been established to restore and create the milkweed habitats on which it depends.

It is important that we do not try to excessively regulate to conserve butterfly populations. We need people to be part of the process and be the power on the ground behind conservation programs.

This will be particularly significant for the common and widespread species now experiencing decline in urbanized western countries. The importance of even small gardens for helping with the conservation of pollinators such as butterflies has been demonstrated by scientific study. So, we can play a role in ensuring that butterflies will exist for posterity by planting butterfly-friendly gardens. We want future generations to experience nature by touching and being part of it, so that they will value the little things that run the world.

EXPLORING THE LIVES OF BUTTERFLIES

This book provides a close-up view of day-to-day butterfly lives, featuring aspects of their life history, behavior and habits, habitats and resources, populations, seasonality, and defense. It also looks at the human-caused threats currently affecting butterflies and what we can do to help them survive and prosper.

→   The iconic Monarch butterfly (*Danaus plexippus*) is an ambassador for butterfly and pollinator conservation. North American populations have declined drastically this century, attracting widespread public concern and numerous habitat restoration programs.

← Tropical swallowtails like this Emerald Peacock (*Papilio palinurus*) are among the largest and most brightly colored butterflies in the world.

↓ The life of an adult butterfly begins when it emerges from the pupa. Within hours it is ready for the first flight and a life that may occupy days, weeks, or months.

Each chapter focuses on a particular topic, delving into it in more detail with examples, feature boxes, and detailed illustrations. Following these discussions are profiles of butterfly species that exemplify the particular aspect of life history being explored. Each of these provides the species' names and taxonomic details, its key characteristics, descriptions of its different life stages, information on its habits and behavior, and unique points of interest. Accompanying these sections are dazzling photos of these magnificent creatures and maps showing where they can be found.

Life Histories introduces the different butterfly families and describes their characteristics. It then explores the lives of butterflies, including their different life stages: egg, caterpillar, pupa, and adult butterfly. In Butterfly Behavior, everyday aspects of butterfly lives are discussed, such as flight behavior, feeding, roosting, territoriality, courtship, mating, and mobility. The varied habitats butterflies occupy and use, along with the importance of climate and food plants, are discussed in Habitats & Resources and Butterfly Populations goes on to explore butterfly abundance, population dynamics, dispersal, and migration.

Butterfly Seasonality focuses on how these insects overcome the challenges of annual weather patterns, while Defense & Natural Enemies investigates the methods and strategies they use to defend themselves. The final chapter, Threats & Conservation, discusses the many and increasing dangers that butterflies face, along with actions that we can take to help the species. Following the main chapters is a glossary of terms used in the book, and a list of useful resources.

LIFE HISTORIES

The butterfly families

Butterflies are classified into seven families based on their evolutionary history, and each of these groups shares physical, behavioral, and ecological features, including body structure, wing characteristics (venation, patterning, and color), host plants, and flight. These families have Latin names as well as common names such as "swallowtails," "skippers," and "brushfoots."

FAMILY HEDYLIDAE
Moth-like butterflies

36 species

NO SUBFAMILIES
One genus: *Macrosoma*

FAMILY HESPERIIDAE
Grass skippers, spreadwing skippers, skipperlings, awls, awlets, policemen, firetips

4,200 species

13 SUBFAMILIES
Barcinae

Chamundinae

Coeliadinae ↓
(awls, awlets, policemen)

Eudaminae

Euschemoninae
(Regent Skipper)

Hesperiinae
(grass skippers)

Heteroptinae (skipperlings) ↑

Katreinae

Malazinae

Tagiadinae

Pyrrhopyginae (firetips)

Pyrginae
(spread-winged skippers)

Trapezitinae

FAMILY PAPILIONIDAE
Parnassians and swallowtails (papilionids)

600 species

THREE SUBFAMILIES
Baroniinae (monotypic, *Baronia*)

Papilioninae ↓ (swallowtails)

Parnassiinae (parnassians)

FAMILY PIERIDAE
Whites, marbles, and sulphurs (pierids)

1,100 species

FOUR SUBFAMILIES
Coliadinae (Sulphurs) ↓

Dismorphiinae (mimic sulphurs)

Pierinae (whites) ↓

Pseudopontiinae

FAMILY RIODINIDAE
Metalmarks (riodinids)

1,500 species

THREE SUBFAMILIES
Euselasiinae

Nemeobiinae ↓

Riodininae

FAMILY LYCAENIDAE
Gossamer wings: coppers, hairstreaks, elfins, sunbeams, harvesters, and blues (lycaenids)

Circa 5,500 species

SEVEN SUBFAMILIES
Aphnaeinae

Curetinae (sunbeams)

Lycaeninae (coppers) ↓

Miletinae (harvesters)

Polyommatinae (blues) ↓

Poritiinae

Theclinae (hairstreaks) ↓

FAMILY NYMPHALIDAE
Brushfoots: milkweed butterflies, fritillaries, admirals, ladies, tortoiseshells, anglewings and commas, buckeyes, checkerspots and crescents, satyrs, browns, ringlets, leafwings, snouts, longwings, and emperors (nymphalids)

6,300 species

12 SUBFAMILIES
Apaturinae (emperors) ↓

Biblidinae

Calinaginae

Charaxinae (leafwings) ↓

Cyrestidinae

Danainae (milkweeds)

Heliconiinae (longwings)

Libytheinae (snouts)

Limenitidinae (admirals) ↓

Nymphalinae

Pseudergolinae

Satyrinae (Browns)

HESPERIIDAE: SKIPPERS

Skippers' wings are generally short and triangular, and their bodies are stout with a wide head. Skippers have the clubs of their antennae hooked backward while other butterflies have swollen tips on their antennae. Many temperate skippers are dark with simple wing patterns, but some tropical skippers are larger and colorful. More than 4,200 species of skippers occur worldwide, with the greatest diversity in Central and South America. Their common name comes from their quick, darting flight.

There are 13 subfamilies of skippers and most species feed on grasses, bamboos, palms, and sedges as caterpillars. Skippers can be difficult to identify on the wing because of their small size, rapid flight, and subtle markings. The smaller grass skippers are known in the USA as "skipperlings," for example the Garita Skipperling (*Oarisma garita*) and the very similar, but introduced, species, European Skipperling (*Thymelicus lineola*).

Skippers have a long proboscis or "tongue" relative to their body length, which they use to suck nectar. Like other butterflies, many skippers perch with their wings folded above their bodies. However, others spread their wings at rest or hold the hind wings out horizontally and the forewings upright, slightly cocked open. Some male skippers have well-defined scent scales on their forewings. They use these to dispense pheromones during courtship to entice females to mate.

← A Long-tailed Skipper (*Urbanus proteus*) probes a flower for nectar with its long proboscis.

→ A Northern Cloudywing (*Thorybes pylades*) basking and showing the hooked antennae that are characteristic of the skipper family.

PAPILIONIDAE: SWALLOWTAILS

Some of the largest and most spectacular butterflies in the world belong to this family. The Queen Alexandra's Birdwing (*Ornithoptera alexandriae*) in New Guinea has a wingspan of 12 in (30 cm). There are around 580 species of swallowtail butterflies worldwide, with the greatest diversity in the tropics. Only 12 species are found in Europe and just one in the UK. North America has 40 species, including the Pale Tiger Swallowtail (*Papilio eurymedon*).

Swallowtails are avid feeders, preferring large flowers. Some keep their wings vibrating to help support their bodies while perched feeding. They are also frequent "puddlers"—gathering in groups to imbibe minerals and salts from damp sand.

Swallowtails are strong fliers and can disperse quite far, including short distances over open water. Some swallowtails have hind wing extensions or "tails" that give them their common name. One subfamily, the parnassians or apollos, live mostly in mountainous areas, lack tails, and are translucent white with black and red spots.

↖ A male Queen Alexandra's Birdwing of New Guinea. The female of the species is the largest butterfly in the world, with a wingspan of up to 12 in (30 cm).

← The Pale Tiger Swallowtail, one of 40 North American species of swallowtails, feeding from a flower and showing the characteristic "tails" of this butterfly family.

↗ A Becker's White butterfly from western North America displaying its intricate green ventral patterning.

→ The Common Jezebel (*Delias eucharis*), which is native to South Asia, takes a sip of nectar.

PIERIDAE: WHITES, MARBLES, AND SULPHURS

The butterfly family Pieridae has about 1,100 species in four subfamilies worldwide. Most pierid butterflies are small- to medium-sized and white, yellow, or orange, often with dark spots. Some have sex-specific ultraviolet patterns used in courtship that are invisible to the naked eye. The word "butterfly" is thought to have been derived from a European member of this family, the yellow-colored Common Brimstone (*Gonepteryx rhamni*), which was formerly known as the "butter-colored fly."

Most of the mimic sulphurs in the subfamily Dismorphiinae are found in the New World tropics, where they resemble toxic clearwing and longwing butterflies. The obscure subfamily Pseudopontiinae has just a handful of African species in the genus *Pseudopontia*, and might be called the paradoxes because they lack antennal clubs. The whites (Pierinae) and sulphurs (Coliadinae) are found around the globe, from seashores to mountain tops. Marbles (*Euchloe* spp.) are mostly found in the northern hemisphere and are often some of the earliest spring butterflies. The Becker's White (*Pontia beckerii*) is another early spring butterfly in western North America with intricate green ventral patterning, contrasting with the mostly white upperside.

Most sulphur caterpillars eat legumes, and whites in temperate zones prefer mustards. However, many tropical species—about one-third of all pierids—feed on mistletoes, which are plant parasites that embed themselves in mature trees. Jezebels comprise the largest butterfly genus in the world with around 250 *Delias* species found throughout Asia and the Australasian regions. The infamous Large Cabbage White (*Pieris brassicae*) and Small Cabbage White (*Pieris rapae*) butterflies can be serious pests, by virtue of their caterpillars feeding on cabbages and other cultivated crucifers.

LYCAENIDAE: BLUES, COPPERS, AND HAIRSTREAKS

Lycaenidae (pronounced ly-SEE-nid-day), the second largest butterfly family, includes about 5,500 mostly small (less than 2 in/5 cm) species worldwide. The family is often referred to as the "blues, coppers, and hairstreaks," but this diverse group includes the sunbeams, gems, harvesters, and others. Therefore, the easiest name to refer to this family is simply the "lycaenids."

The adults are typically small to tiny and they comprise about 30 percent of all known butterfly species. Coppers are especially dominant in northern temperate regions, blues are richest in the Old World tropics and northern temperate zones, and hairstreaks are particularly abundant in the New World tropics.

The Brown Elfin (*Callophrys augustinus*) is in a subgroup of hairstreaks called "elfins" that usually lack the tiny tails found on the hind wings of most hairstreaks.

Adult hairstreaks and blues often have antenna-like tails on their hind wings that they slowly move around while at rest to confuse predators, making it difficult for them to determine which end is the head.

Many lycaenid caterpillars and pupae associate with ants in one way or another. Caterpillars are attended by ants sharing nutritious secretions from special glands in exchange for protection from enemies. Many lycaenid caterpillars have unusual diets and feed on lichens, cycads, ant larvae, or aphids and other soft-bodied insects.

←← The flimsy hind wing tails of this male Powdered Oakblue (*Arhopala bazalus*) twist in opposite directions. This causes them to move up and down alternately with the slightest breeze. These fake "antennae" may fool predators into attacking the wrong end, allowing the butterfly to escape.

← The Brown Elfin (*Callophrys augustinus*), a spring butterfly of western North America. Elfin butterflies are hairstreaks without tails.

→ The Mormon Metalmark (*Apodemia mormo*) flies in late summer and is the only metalmark found in the northern part of North America.

RIODINIDAE: METALMARKS

Most of the more than 1,500 metalmark butterflies live in the tropics, but there are a few in more temperate regions, such as the Mormon Metalmark (*Apodemia mormo*), which extends from Mexico as far north as Canada. Metalmarks occur in a variety of habitats, but the tropical rainforests of South America are the center of their diversity.

Metalmarks are small- to medium-sized butterflies with short, stocky wings and long antennae. Small metallic spots on the wings of most species give these butterflies their common name.

Many metalmarks are brilliantly colored, like the Blue Metalmark (*Lasaia sula*) in central America or the Common Red Harlequin (*Paralaxita telesia*) in

Asia. The enigmatic, charcoal-black *Styx infernalis* was first classified as a moth, then reclassified in Pieridae, Lycaenidae, or the only member of its own family. Genetic evidence has established that it is the descendant of an Asian metalmark lineage that crossed the Bering Strait long before humans and left no descendants in North America. Like lycaenids, some metalmark caterpillars associate with ants, but their specialized organs for feeding ants and communicating with them are different.

NYMPHALIDAE: BRUSHFOOTS

Nymphalidae is the largest family of butterflies. With more than 6,300 species worldwide, it shows enormous morphological complexity and diversity. They are medium- to large-sized butterflies and are often strikingly patterned and colored. Many brushfoots are skilled aviators, most are fast-flying and elusive, and a few undergo annual transcontinental migrations.

Like most butterflies, the diversity of species in this family is strongest in the tropics. Many tropical species are brilliant and colorful, such as the morphos (*Morpho* spp.) in the New World, *Euphaedra* spp. in Africa, and the Glorious Begum (*Agatasa calydonia*) in Asia. Wing shape is highly variable, from irregular margins to tail-like projections.

Nymphalids use only the four rear legs for walking; the first pair of legs are rudimentary, covered with brushlike "fur," and dedicated to sensory functions. Brushfoot coloration is complex and varied, drawing on a wide palette of colors, from reds and browns to black, yellow, silver, greens, and blues. The longest-lived North American butterflies are all nymphalids, with many species passing the winter as adults.

→ The colorful Glorious Begum (*Agatasa calydonia*), is restricted to undisturbed rainforests in part of Southeast Asia.

↘ Most moth-like butterflies, such as this *Macrosoma* sp. from Belize, fly at night, lack antennal clubs, and listen for bats with ears located on their wings.

↓ The Blue Morpho (*Morpho helenor*) is a spectacularly iridescent brushfoot butterfly found in the tropical rainforests of South America.

HEDYLIDAE: MOTH-LIKE BUTTERFLIES

These butterflies fly at night and lack the swollen antennal tips typical of other butterfly families. Until recently, they were classified as moths. Genetic data affirm that these are indeed nocturnal and crepuscular butterflies most closely related to skippers. The adult butterflies have hearing organs on their wings to detect bat ultrasound. The family comprises just over 30 species in a single genus, *Macrosoma*, found only in Central and South America. The larvae feed on a variety of different plants, including *Croton* spp. and *Theobroma* spp.

Is it a butterfly or a moth?

Butterflies and moths form the order of insects called Lepidoptera, which means "scale winged." Human beings like to separate butterflies from moths, but in fact butterflies are just one group of moths specialized to fly during the day.

THE ANTENNAE ARE USUALLY THE KEY

Is it a butterfly or is it a moth? This is a question that has no easy answer. However, it is the most common question that people have when they find an insect with four colorful wings. If the creature is bright and colorful and flying in the daytime, the questioner will usually guess that it is a butterfly. However, some people are surprised to learn that many moths are also decked out in bright colors and fly by day as well.

Consequently, color and daytime activity are not good identifying features for a butterfly. Neither is visiting flowers, because both butterflies and moths are important daytime pollinators. However, most butterflies do not fly at night. The majority of butterflies rest with their wings closed above their body as opposed to resting with wings laid flat as most moths do, but there are some exceptions.

The best identifying feature for a butterfly is its antennae. Butterfly antennae are almost always clubbed at the end. The most prominent exceptions, moth-like butterflies and paradoxes, have already been mentioned. Some moth families, such as the giant butterfly moths (family Castniidae), have clubbed antennae, but these are uncommon. Most

↑ The Sagebrush Sheep moth (*Hemileuca hera*), an attractive day-flying moth in western North America, is often mistaken for a butterfly.

moths have thin, wiry antennae, while some larger moths have elaborate, feathery antennae.

Lepidopterists appreciate both butterflies and moths, but in reality most have a strong preference for one group or the other. Butterfly enthusiasts consider some of the more interesting moth species to be "honorary butterflies."

THE ANATOMY OF A BUTTERFLY

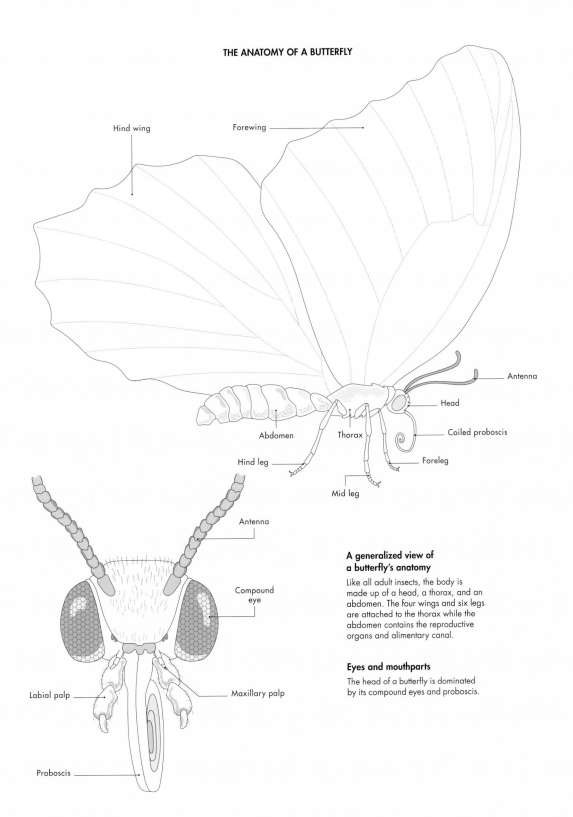

Hind wing

Forewing

Antenna

Head

Coiled proboscis

Abdomen

Thorax

Foreleg

Hind leg

Mid leg

Antenna

Compound eye

Labial palp

Maxillary palp

Proboscis

A generalized view of a butterfly's anatomy

Like all adult insects, the body is made up of a head, a thorax, and an abdomen. The four wings and six legs are attached to the thorax while the abdomen contains the reproductive organs and alimentary canal.

Eyes and mouthparts

The head of a butterfly is dominated by its compound eyes and proboscis.

Eggs and oviposition

Female butterflies usually lay (oviposit) a few hundred or up to 1,000 eggs during their lifetime, although only a few females manage to survive long enough to lay their full complement of eggs. Far more eggs are laid than will survive to be adult butterflies. Eggs and caterpillars suffer extremely high mortality from natural enemies such as predators, parasitoids, and pathogens as well as from non-biotic factors such as bad weather and food shortages.

WHERE BUTTERFLY LIFE BEGINS

Female butterflies have a pair of ovaries, and each ovary contains four ovarioles or tubes in which eggs form and mature as they move down the oviduct to the ovipositor, or egg-laying tube, to be laid. Some butterflies eclose (emerge) from their pupa with mature eggs ready to be laid soon after mating. Other species need a few days of adult feeding and warmth to mature their eggs. Mating and egg-laying are the primary functions of every female butterfly.

A KALEIDOSCOPE OF EGGS

Butterfly eggs are naturally very small, but there is a tenfold range, from the minuscule to the relatively robust. The Pygmy Blue (*Brephidium exilis*), one of the smallest butterflies in the world, lays eggs that are just 0.2–0.4 mm in size, while those of the world's largest butterflies, the birdwings (Papilionidae), may have a diameter of 4 mm. However, most butterfly eggs are about 1 mm in diameter. Many are white, green, or bluish and blend in with their surroundings, but a few—particularly of species in the family Pieridae (including the orange-tips and sulphurs)—become orange or bright red as they mature.

Butterfly eggs have a chitin-based shell and are fragile, making them easily punctured or damaged. They are an attractive food or host for many kinds of small insect predators and parasitoids (parasites that kill;

see pages 234–237). Butterflies will often lay their eggs singly on the lower surface of leaves, out of the sun and away from the eyes of many predators. Red Admirals (*Vanessa atalanta*), Small Cabbage Whites (*Pieris rapae*), and swallowtails (Papilionidae) do this, while other species such as tortoiseshells (*Aglais* spp.) and Common Imperial Blues (*Jalmenus evagoras*) will lay their eggs in large batches that may contain hundreds. Other species, including the European Skipperling (*Thymelicus lineola*), will fall somewhere in the middle, laying strings or small groups of up to a dozen eggs.

Butterfly eggs may be smooth or highly ornamented, and they may be flattish or bottle-shaped. Some have projections and most are ribbed in some way. Such is the variety of ornamentation that many eggs can be identified to species level under the microscope. The shape of the egg is usually a guide to the butterfly family to which it belongs. White butterflies (Pieridae) usually have bottle-shaped eggs, while the blues, coppers, and hairstreaks (Lycaenidae) have eggs shaped like miniature sea urchins or buttons. The ornamentations are thought to regulate humidity. Many butterfly eggs are susceptible to desiccation. At the top of the egg is an opening called a micropyle, through which sperm enters as the egg leaves the female abdomen.

REPRODUCTIVE SYSTEMS OF A FEMALE AND MALE BUTTERFLY

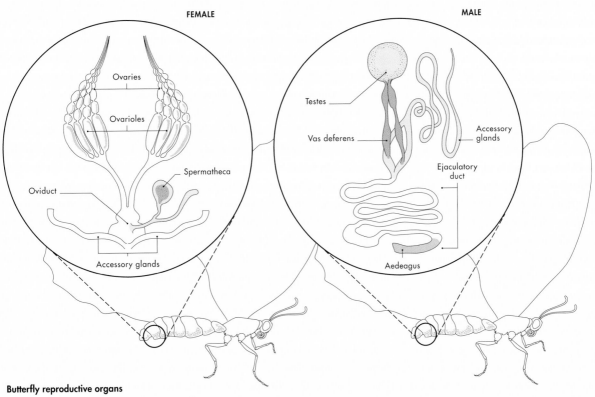

FEMALE

Ovaries

Ovarioles

Spermatheca

Oviduct

Accessory glands

MALE

Testes

Vas deferens

Accessory glands

Ejaculatory duct

Aedeagus

Butterfly reproductive organs

The reproductive systems of male and female butterflies occupy much of the abdomen. The male system is based on a pair of sperm-producing testes that are usually held together in a single sac. The female system is dominated by the formation of immature and mature eggs that occupy the ovarioles, causing the abdomen to be noticeably plumper than the male abdomen.

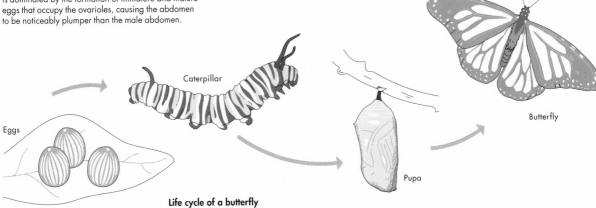

Caterpillar

Butterfly

Eggs

Pupa

Life cycle of a butterfly

A female Monarch butterfly lays up to 800 eggs on milkweeds during its lifetime. The caterpillar develops through five instars, then forms a hanging jade-green pupa where it transforms into an adult butterfly. Development from egg to adult takes about four weeks during the summer.

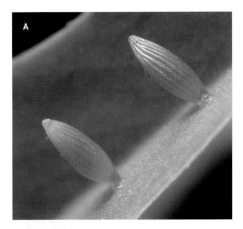

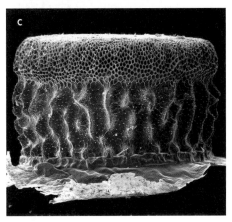

CHOOSING THE RIGHT SPOT

Once mated, and carrying mature eggs, females search for the host plants that will feed their caterpillars. Some butterflies use only a single host plant, others specialize on a single plant family, and yet others use a wide range of host plants. The Monarch (*Danaus plexippus*) is confined to milkweeds (Apocynaceae) as host plants, while the Painted Lady (*Vanessa cardui*) uses plants in many families as food for its caterpillars. The rare Leona's Little Blue (*Philotiella leona*), found only in a small area of high desert in Oregon, USA, has but a single host plant, the similarly rare spurry buckwheat (*Eriogonum spergulinum*).

Females locate the correct host plants using both sight and smell, and after much inspection will alight and "taste" the leaf surface with their "feet." If the plant is the right one and is in good condition, she will lay an egg or batch of eggs. Females are rarely satisfied with the first plant or plant part they test, and tend to visit a number before finally bending their abdomen and laying an egg.

↖ A) The eggs of pierid butterflies (whites and sulphurs) are bottle-shaped and pale-colored when first laid but turn orange-red with maturity. B) The eggs of duskywings (Hesperiidae) are strongly ribbed and often boldly colored yellow, red, or orange. C) The eggs of *Liphyra brassolis* (Lycaenidae) are flattened and button-shaped with finely or coarsely reticulated surfaces.

←← A Map (*Araschnia levana*) butterfly prepares to lay an egg on the underside of a leaf.

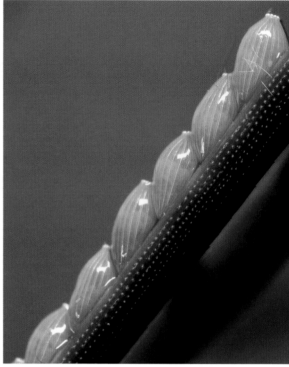

A female's choice of host plant is critical to the future of her progeny. The plant must offer food in sufficient quantity and quality for the developing caterpillars. Females will often reject seemingly suitable plants, presumably because they detect some chemical or physiological deficiency. Some butterflies—particularly blues and hairstreaks (Lycaenidae)—choose to lay their eggs on the parts of plants that are higher in nitrogen, such as buds and flowers. The caterpillars then feed on these nitrogen-rich plant parts, although they can make do with leaves if necessary.

The decision made by a female butterfly in choosing a plant and a location on that plant for an egg is one of the most important moments in a butterfly's life. It has a huge bearing on the success of the caterpillar, which must live out its mother's choice. Fortunately, female butterflies have been fine-tuned through evolution to make the best choices possible for their offspring.

While some butterfly species, such as the Silvery Blue (*Glaucopsyche lygdamus*), lay their eggs on buds or flowers, other species, including some brown butterflies in the family Nymphalidae, simply scatter them as they fly over host plants. Most butterflies lay

↖ The female Mourning Cloak (*Nymphalis antiopa*) spends a few hours carefully arranging her ribbed, orange eggs in a collar around a willow twig. She may lay up to 200 eggs in this manner.

↑ Flask-shaped, blue-green eggs of the western North American Pine White (*Neophasia menapia*) butterfly are laid in angled row batches of 3 to 25 along needles of conifer host plants.

↗ The Western Pine Elfin (*Callophrys eryphon*) female lays her eggs on the soft, new-growth branch tips of pine trees, often deep within crevices.

their eggs on the leaves of their host plants, mostly on the lower surface, but some, such as the Two-tailed Tiger Swallowtail (*Papilio multicaudata*; see pages 54–55), lay them on the upper, exposed surface. Others, such as some fritillaries (Nymphalidae), attach their eggs to ground detritus or tree trunks close to where host plants will appear in a later season.

Butterfly eggs are rarely found without a dedicated search. Knowing where the female typically lays her eggs aids the hunt, so it is "simply" a matter of searching through vegetation to find them. However, given the typical large amount of vegetation and small numbers of eggs, finding them is never easy. Imagine searching through a field of grass for the eggs of Ringlets (*Aphantopus hyperantus*) or Meadow Browns (*Maniola jurtina*)! Egg searching is easier with a species such as the Mourning Cloak (*Nymphalis antiopa*, see pages 198–199) butterfly, which lays its eggs in large masses encircling twigs of willows.

THE DEVELOPING EGG

Butterfly eggs develop rapidly—usually within a week or so, but more quickly when conditions are warm and slower when they are cool. Sometimes, eggs are programmed to delay development, spending periods of adverse weather conditions such as extreme heat, cold, or drought in a state of arrested development known as diapause, or estivation. They hatch when optimal conditions return. The ability of diapausing butterfly eggs to withstand adversity is remarkable. For example, the grenade-shaped eggs of the Pine White (*Neophasia menapia*) are laid in neat rows along pine needles and survive freezing temperatures for many months in the American Northwest before hatching in the spring.

Caterpillars: eating machines

Young butterflies are called larvae (singular: larva) or caterpillars. Once they hatch, they turn into veritable eating machines, using the energy from their food to produce the building blocks and mainframe that will eventually become the adult butterfly.

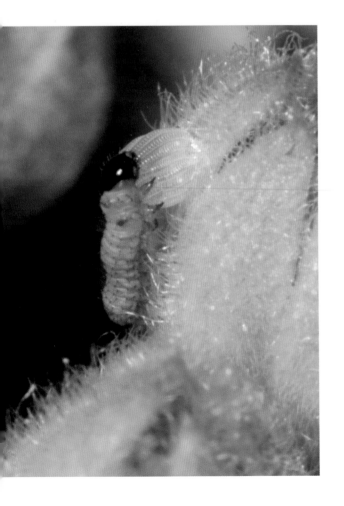

RAPID GROWTH

Caterpillar lives are just as varied and complex as butterfly adult lives—arguably more so. Once a caterpillar hatches from its tiny egg, it has a lot of feeding and growing to do. Caterpillars may increase their mass by more than a thousand times before they reach pupation, and they need to do this quickly. The more time they spend as a caterpillar, the more exposure they have to the many natural enemies that are searching for a juicy meal. Some caterpillars can complete their development in as little as eight days, while a few temperate species living in areas with short summers may take more than a year.

← The first meal of a newly hatched caterpillar is often its own eggshell. Some species eat the entire shell, others eat it partially, while other species may not eat it at all.

→ A newly molted caterpillar of the Oregon Swallowtail (*Papilio bairdii*). The shed skin can be seen behind the caterpillar, which is now in its final instar.

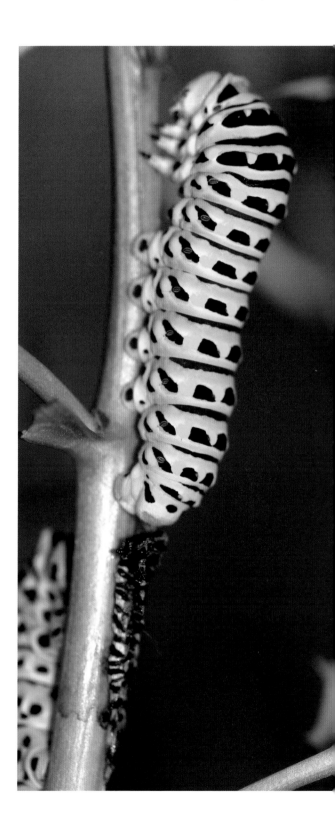

CONSTRAINTS AND PERILS

All caterpillars have a skin, or cuticle, that tightens with growth and needs to be replaced with a larger size. They do this by molting—at least four times, and in some species up to ten times. Caterpillars usually double in length between each molt, but most growth in terms of bulk occurs in the final instar, or stage. During the final molt, the caterpillar becomes a pupa (plural: pupae), sometimes called a chrysalis.

When an adult butterfly hatches out of its pupa, or ecloses, it has already run the gauntlet of natural enemies and adverse environmental conditions that usually results in a less than 10 percent survival rate for eggs, caterpillars, and pupa. The caterpillar's natural enemies include birds, reptiles, amphibians, mammals, other insects, and pathogens. However, caterpillars put up a good fight and have evolved many defense strategies, including concealment, evasion, threats, scare tactics, chemical deterrents, aggregation, and the use of bodyguards. Utilizing one or more of these tactics allows enough immature butterflies to survive and perpetuate the species.

HIDING

Concealment is probably the defense strategy most frequently used by caterpillars, and the reason it is often hard to find them. Some caterpillars hide themselves within their host plants or even burrow into buds and seeds. Many caterpillars feed by night and hide by day at the base of their food plants. Others hide within leaves that they stitch together with silk to create a tent. Camouflage or crypsis is an effective strategy for diurnal (daytime) feeders, and many leaf-eating caterpillars are green, helping them to blend into the foliage. Those that feed on flowers may match the red, yellow, or white markings of their floral food. Grass-feeding caterpillars are usually a shade of green with darker or paler stripes, which effectively hides them in a sea of grass stems and blades.

When concealment fails and a caterpillar is found by a predator, it may show aggression, including sudden jerking movements to scare the attacker. This tactic is most effective when performed in unison by a large group of caterpillars, as demonstrated by some brushfoot species, such as the California Tortoiseshell (*Nymphalis californica*). Some late-instar caterpillars display and move their mandibles as if to bite. However, these aggressive behaviors are empty threats, and no caterpillar can actually harm a predator.

↖ A caterpillar of the Propertius Duskywing (*Erynnis propertius*) resting in a shelter it has made by silking down the edge of the leaf to create a bivouac.

← This trio of Common Palmfly caterpillars (*Elymnias hypermnestra*) rest on a palm frond of their host plant.

↗ The Small Cabbage White (*Pieris rapae*) is one of the few butterfly species that can cause economic damage by feeding on cultivated crucifers such as cabbage and cauliflower. Here, young caterpillars are feeding on a flower head of broccoli.

UNWARRANTED BAD PRESS

A major reason that caterpillars get bad press is that some are significant pests of agricultural crops and garden plants. However, the vast majority of these "pests" are actually moth caterpillars, not butterfly caterpillars. The most significant butterfly pests are the European Small Cabbage White and Large Cabbage White (*Pieris rapae* and *P. brassicae*, respectively), well known to gardeners of cruciferous vegetables. In Asia, the Guava Blue (*Deudorix isocrates*) is a pest of cultivated fruit including pomegranate and guava. But even these rarely destroy entire crops. Most butterfly caterpillars cause no visible or economic damage to ornamental or crop plants, and the vast majority use native species as their food plants rather than exotic crop or garden species.

EXPOSED

Other caterpillars take the opposite approach to concealment, resting openly with gaudy and striking coloration so predators cannot miss them. However, the colors—usually contrasting hues such as black and white or black and yellow—are a warning that says, "Eat me, and you'll regret it!" These caterpillars contain toxins (usually sequestered from their host plant) that will cause vertebrate predators to fall ill and even die. The most famous example of this is the Monarch (*Danaus plexippus*), whose black-, yellow-, and white-banded caterpillar incorporates cardiac glycoside toxins from milkweed to defend itself, primarily from birds. So successful is this strategy that caterpillars of other non-toxic species will mimic the Monarch caterpillar's coloring to gain some protection against "once bitten" predators.

↖ The aposematic black, white, and yellow caterpillar of the Indra Swallowtail (*Papilio indra*). These striking caterpillars advertise their distastefulness to educated birds who avoid taking them as food.

↑ Early-instar caterpillars of some species, such as these California Tortoiseshell caterpillars (*Nymphalis californica*), aggregate for protection.

↗ Caterpillars of the Pale Imperial Hairstreak (*Jalmenus eubulus*) and its relatives provide essential nutrients to Meat ants (*Iridomyrmex* spp.) through various glands. In return, the ants protect the butterflies from predators and parasites throughout their immature stages.

Aggregation or gregariousness is another caterpillar behavioral tactic that reduces the odds of any single individual being attacked. This is often used by early-instar caterpillars before being superseded by other defense tactics as they mature. Communal caterpillars may also build silken webs, supports, and platforms to help keep the community together and dissuade natural enemies.

THE BENEFITS OF MUTUALISM
Some caterpillars, and particularly blues, coppers, and hairstreaks (Lycaenidae), have developed a defense strategy based on recruiting ant bodyguards to repel threats from predators and parasitoids.

Ants may be a significant predator of some butterfly caterpillars, but not the lycaenids, which have small glands that, when stimulated, excrete a sugar-rich "honeydew" that ants love. In a clear example of mutualism, the ants are provided with a nutritious food supply and the caterpillars receive protection from ants crawling over and around them, effectively preventing attacks from parasitic wasps, predatory bugs, spiders, and many other natural enemies.

Pupae: crucibles of transformation

The pupa, also known as a chrysalis, is the transformative stage between the hungry caterpillar and adult butterfly. The egg and the pupa are the only life stages of a butterfly that are immobile, although some pupae can flex if disturbed.

FORMS OF PUPAE

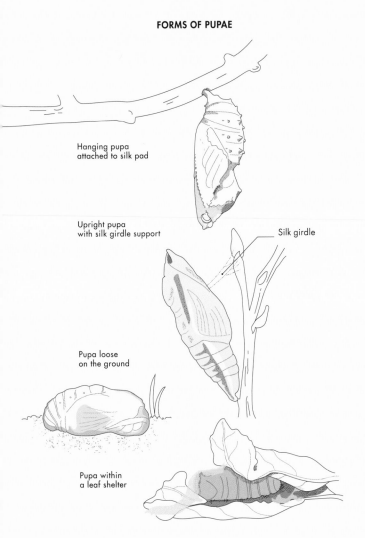

Hanging pupa attached to silk pad

Upright pupa with silk girdle support

Silk girdle

Pupa loose on the ground

Pupa within a leaf shelter

PUPA MODES

Butterfly pupae develop in one of four basic modes: loose, or sometimes in a sparsely spun cocoon on the ground; within a leaf shelter; hanging by the terminal end that is attached to a silk pad; or attached upright with a supporting silk girdle.

Skipper butterflies (Hesperiidae), including the European Skipperling (*Thymelicus lineola*; see pages 56–57), commonly form pupae in tied leaf or grass shelters, while hanging pupae are characteristic of brushfoot butterflies such as the Red Admiral (*Vanessa atalanta*; see pages 58–59). Girdled pupae are most often found in species in the families Papilionidae, Pieridae, and Lycaenidae. Forming a pupa that is either loose or lightly cocooned on the ground is rare in butterflies, although it is common in moths. Parnassian butterflies (members of the swallowtail family, Papilionidae), including the Mountain Parnassian (*Parnassius smintheus*; see pages 98–99) in western North America, eclose from pupae formed within a sparsely spun cocoon on the ground, as do a few species in the brushfoot family (Nymphalidae).

TRANSFORMATION

Once a pre-pupal caterpillar has selected its pupation site, it shrinks a little and waits motionless for a day or so for the final molt.

Caterpillars that form hanging pupae adopt a characteristic "J" shape. After 24–48 hours, the caterpillar's skin splits behind the head, revealing the soft pupal case that has formed beneath. With much wriggling, the caterpillar skin moves down the body, revealing more of the soft new pupa. Once the shed skin reaches the last segment, the tip (also known as a cremaster) probes and seeks the silk pad spun by the pre-pupal caterpillar. With hanging pupae, this is a critical phase; if the cremaster fails to make contact with the silk pad, the soft and vulnerable pupa will fall to certain death. After attachment, more wriggling occurs, which results in the shed skin falling away. After a few minutes the pupa stops moving, hardens, and becomes the color that protects it until the adult stage.

J-shaped pre-pupal caterpillar

A typical pre-pupal caterpillar before transforming into the pupa or chrysalis.

Ecdysis: from pre-pupal caterpillar to pupa

The process of a caterpillar shedding its skin for the final time and transforming into a pupa is called ecdysis. This swallowtail caterpillar is shown in pre-pupal, mid-ecdysis, and new pupal stages.

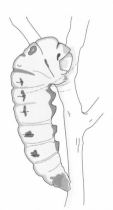  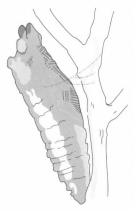

The pre-pupal stage of a caterpillar awaiting its final molt.

The skin moves down the caterpillar's body as it molts.

The caterpillar's skin is fully molted, leaving a soft pupa prior to it hardening.

HIDDEN FROM VIEW

Being immobile and unable to defend itself, the pupa must remain hidden in a protected location away from foraging predators and parasitoids. This is invariably the case for most butterflies, the pupa being the least-seen stage in its life cycle. In its last day or two, the full-grown caterpillar has the crucial task of finding a suitable location for its transformation into a pupa. Most caterpillars at this point become duller in color and begin to wander in search of the perfect place to pupate. Some, such as the Red Admiral, will pupate on their host plant but will bind together some leaves with silk to create a "tent" to cover and hide the pupa. For a wandering caterpillar, the perfect location may be a twig within a bush, a blade of grass within a meadow, or simply on the ground. It may travel up to 330 ft (100 m) from its host plant in search of such a spot.

In addition, pupae are cryptically colored and shaped. For example, the pupae of orange-tip (*Anthocharis* spp.) butterflies are colored to match their background and are shaped like a large thorn. The pupae of butterflies defended by toxins, on the other hand, are often showy to advertise their distastefulness. The pupae of crow butterflies (*Euploea* spp.), for example, look like molten metal. Undoubtedly, some pupae still fall prey to foraging birds and small mammals, and parasitic wasps can locate them by smell. For most species, however, if a caterpillar is able to transform into a pupa and manages to escape from parasitoids, there is a good chance it will produce a butterfly.

THE MAGIC WITHIN

The pupa is the least-studied stage of butterfly life, yet it is the most transformative. We still do not know exactly what happens during this process, but modern imaging techniques are allowing glimpses. Micro X-ray computed tomography scans now allow us to see some of the development that happens within a pupa. For pupae that develop during the spring or summer, this remarkable transformation takes just a week or two. For many species—including orange-tips (Pieridae), swallowtails (Papilionidae), and hairstreaks (Lycaenidae)—the pupa is the overwintering stage, so it can remain dormant for months or, in some cases, two or three years.

LATERAL VIEW DAY 1

LATERAL VIEW DAY 7

LATERAL VIEW DAY 13

↑ Development within the pupa of a Painted Lady (*Vanessa cardui*) butterfly over 13 days as shown by high-resolution X-ray computed tomography scans. Red shows the forming gut and blue shows the tracheal (breathing) system. The Malpighian tubules, which function like a kidney, are shown in orange.

← Pupae of grass-feeding caterpillars like those of the Common Wood Nymph (*Cercyonis pegala*) in North America are beautifully camouflaged in a sea of grass.

A BUTTERFLY IS BORN

A few days before an adult butterfly emerges, or ecloses, the pupa darkens. On the day before, the pupal shell becomes transparent, showing the patterns and color of the butterfly's wings. At this stage, eclosion is just hours away. Most butterfly pupae eclose soon after sunrise to optimize successful emergence, post-eclosion drying of wings, and inaugural flight.

Eclosion begins with the butterfly pushing with its legs against the pupal shell. Once the legs are free, the butterfly uses them to grab hold of the shell and pull the rest of itself out. The newly eclosed butterfly then hangs from the shell or nearby support as its initially tiny, crumpled wings slowly increase in size as hemolymph, or insect blood, is pumped through their veins. This is a vulnerable time for a butterfly, as it is unable to fly away from predators and its wings are easily damaged. All being well, the wings will reach their full size within 15 minutes, although full hardening may take a few to several hours depending on temperature. Caterpillars of the Moth Butterfly (*Liphyra brassolis*; see pages 94–95) pupate and eclose inside weaver ant nests and must flee quickly to avoid being eaten. Their crumpled wings remain soft for hours to allow them time to escape and find a safe place to hang their wings to dry.

← Just a few hours from eclosion, this pupa of a Satyr Comma (*Polygonia satyrus*) shows the wing colors of the adult butterfly through the transparent pupa shell.

A METAPHOR FOR CHANGE

The pupa arguably hosts the most incredible "magic trick" in the natural world. The transformation from non-flying, wormlike, hungry caterpillar to colorful, flying adult butterfly is truly one of nature's marvels that will never fail to amaze and make people think. Indeed, this metamorphosis has long been used by people as a metaphor for change, specifically how study of butterflies has been introduced to at least one prison: by rearing Monarch butterflies, inmates were able to recognize and visualize the transformation they could make in themselves. The humble pupa is a powerful symbol for change.

Adult butterflies: beating the odds

The adult butterflies flying around your backyard are success stories, having beaten the overwhelming odds against their survival. They are among the few survivors in a population that have avoided death, the threat of which is omnipresent throughout their immature lives.

SURVIVAL OF THE FITTEST AND LUCKIEST

Newly eclosed butterflies have escaped predation, parasitism, disease, and death from unfavorable environmental conditions, including excessive heat, drought, cold, storms, and food shortages. Few butterfly eggs survive to become adult butterflies: in most species the figure stands at less than 10 percent, although often it is much lower than that. The life history of butterflies is truly a story of survival of the fittest and perhaps also the luckiest.

HARBINGERS OF SPRING

After the gloom and cold of winter in temperate regions, nothing lifts the human spirit as much as seeing a butterfly flitting among the first flowers of the year. Usually, the first butterfly to appear will be a species that has hibernated and been awoken by the first warm day of early spring. Many people are stunned to know that an adult butterfly (also called an imago) can overwinter, withstanding often severe cold periods, to emerge at the first hint of spring. In Europe and North America, early-spring butterflies include nymphalids and brushfoots such as a tortoiseshell or Red Admiral (*Vanessa atalanta*).

A little later, the true harbinger of spring on these continents appears: the orange-tips (*Anthocharis* spp.), which spend winter as a pupa. Orange-tips are often seen sailing down pathways and wooded lanes, checking out crucifers for nectar and egg-laying. In many parts of the world the first spring butterfly will be the slightly pestiferous Small Cabbage White (*Pieris rapae*), which also overwinter as a pupa. Even cabbage growers may feel joy at seeing the year's first white butterfly!

↑ The Red Admiral is a welcome sight in spring in Europe and North America, having overwintered or migrated from more southerly climes.

← A Monarch butterfly eclosing from its pupa. With its wings soft and crumpled, this is a very vulnerable time in the life of a butterfly. After a few hours, the wings are firm and straightened and it is ready for the maiden flight.

ANTHOCHARIS JULIA

Julia Orange-tip

Harbinger of spring

SCIENTIFIC NAME	*Anthocharis julia* (W.H. Edwards, 1872)
FAMILY	Pieridae
NOTABLE FEATURES	Orange tips to the forewings
WINGSPAN	1–1½ in (25–38 mm)
HABITAT	Open habitat, particularly lanes, roadsides, meadows, glades, and canyons

The Julia Orange-tip is a true harbinger of spring in western North America. Other orange-tip species are also early spring fliers in other parts of the world. Orange-tip butterflies are unmistakable, with orange corner patches on the forewings.

Orange-tips resonate with people because of their simple yet striking beauty and their welcome early-spring appearance after a long winter. The lower surfaces of the wings are also beautiful, with green marbling and yellow veins. The single annual generation flies for just two or three weeks and spends most of the year as a dormant pupa. The Julia Orange-tip is an avid flower visitor, seeking nectar from spring-flowering plants such as mustards, fiddlenecks, and phlox.

Females lay their eggs singly—usually only one per mustard plant unless plants are scarce, when overcrowding can occur. The eggs are slender, initially white and then turning bright orange-red, and are laid on all parts of the plant. The highly cryptic green caterpillars preferentially feed on buds and flowers before consuming leaves and stems. Blending in with the exact tones of mustard greens helps caterpillars evade detection by predators, although some fall prey to parasitoid wasps and disease.

HEDGING THEIR BETS

In marginal arid environments, some individual Julia Orange-tips may hedge their bets by taking an extended rest as pupae. The pupae are green when formed and the host plant is still green, but turn brown when the plant and those nearby plants wither. While most pupae produce butterflies the following spring, some remain dormant for two or more years. This is thought to spread the risk of butterflies eclosing into unfavorable conditions for survival and reproduction, as can occur when a dry spring causes predominantly drought-stressed or limited numbers of host plants.

→   The Julia Orange-tip is an unmistakable spring butterfly with its namesake orange-tipped forewings contrasting gorgeously with the white hind wings and black body.

Gyas Jewelmark

A brilliant butterfly with an unusual diet

SCIENTIFIC NAME	*Sarota gyas* (Cramer, 1775)
FAMILY	Riodinidae
NOTABLE FEATURES	Blue stripes appear metallic; legs and wing margins covered with long, hair-like scales
WINGSPAN	½ in (10–10.5 mm)
HABITAT	Rainforests from sea level to 800m

The Gyas Jewelmark and its relatives are strikingly colored with bold orange and yellow wings punctuated by shiny blue stripes. The caterpillars' unusual habits are even more interesting.

Unlike some other tropical metalmarks, the caterpillars of the Gyas Jewelmark do not live with ants. In fact, they are covered with long hairs called setae that are directed at an ant or other predatory insects in defense. Upon contact with the hairs, the ants retreat and groom themselves in an apparent attempt to remove small, broken fragments of the minutely barbed setae. The adult butterflies are usually active in the morning, encountered along streams and forest edges perched on leaves. Males stake out a territory and guard it against intruding males, and challenges to their domain by rival males will be contested in a fast-paced, spiraling aerial dog-fight. Those that are successful have a chance of mating with females, which fly along inspecting the males and their territories.

LIVER(WORTS) FOR DINNER

The Gyas Jewelmark caterpillars have an unusual diet. Instead of feeding on vascular plants, as most butterflies do, they feed on liverworts in the family Lejeuneaceae. These ancient plants are related to mosses and grow as epiphylls—tiny plants that live on the leaves of the trees in rainforests. Given the choice between fresh, tender leaves and old leaves encrusted with epiphylls, the caterpillars will always eat the epiphylls, leaving the new foliage untouched. Adults have a more typical diet of floral nectar, often from species of *Alibertia* or *Croton*. They have also been observed feeding on the extrafloral nectaries of some plants. These are small nectar-producing glands found on plant leaves; the nectar is often consumed by ants that guard the plant, protecting it from herbivores.

→ The bedazzled wings of a male Gyas Jewelmark, fringed with hair-like scales, glisten as he rests on the underside of a leaf.

PAPILIO MULTICAUDATA

Two-tailed Tiger Swallowtail

Successful in suburbia

SCIENTIFIC NAME	*Papilio multicaudata* (W.F. Kirby, 1884)
FAMILY	Papilionidae
NOTABLE FEATURES	Large, with yellow and black "tiger" stripes
WINGSPAN	5 in (130 mm)
HABITAT	Canyons, gardens, parks, shrublands, and watercourses

The Two-tailed Tiger Swallowtail is one of the most common large butterflies in North America. In fact, it is the largest butterfly in western North America, with a wingspan of 5 in (130 mm). It can be found in most habitats from sea level to the mountains.

The large size and striking yellow and black stripes of the Two-tailed Tiger Swallowtail catch the attention of even casual observers, especially since the species is common in urban parks and gardens. The single brood flies from late April to August, sometimes joined by a partial second brood late in the summer.

Male Two-tailed Tigers range widely as they seek females, patrolling up and down corridors such as canyons and creek beds. Females spend most of their time in proximity of their host plants or visiting flowers. Two-tailed Tiger Swallowtails use species of *Prunus* (including cherries and plums), ashes, and serviceberries as host plants, many of which are grown as garden ornamentals, helping to explain the success and abundance of this species in urban and suburban areas. Both sexes visit a wide variety of ornamental and native flowers, and males commonly visit mud puddles to gain moisture and minerals, often with other swallowtails.

After mating, the female flies from tree to tree, being very choosy in her search for an oviposition site, before laying a single round green egg on the upper surface of a leaf. She then flies to a different plant—often a considerable distance away—to search for the next oviposition site. Two-tailed Tiger Swallowtails are often on the wing, which is another reason they tend to get noticed.

FAMILIAR CATERPILLARS

Two-tailed Tiger caterpillars are large and green, with false eyespots near the head that the caterpillar enlarges by stretching when threatened. A threatened caterpillar will also produce a malodorous Y-shaped fleshy organ, known as an osmeterium, from behind its head to repulse the threat. Fully fed caterpillars turn dark brown and wander extensively to find a pupation site, often attracting the attention of children. Pupae are formed on upright surfaces, including walls and tree trunks, and blend in with the background. Adult butterflies emerge the following spring.

→ The Two-tailed Tiger Swallowtail, with its five-inch wingspan and contrasting black and yellow "tiger" stripes, attracts a lot of attention as it flies and glides through suburban gardens in the towns and cities of western North America.

European Skipperling

A successful colonizer

SCIENTIFIC NAME	*Thymelicus lineola* (Ochsenheimer, 1808)
FAMILY	Hesperiidae
NOTABLE FEATURES	Small, fast, unmarked skipper
WINGSPAN	¾–1 in (20–25 mm)
HABITAT	Hayfields, meadows, pasture, and waste ground

The European Skipperling, also known as the Essex Skipper in the UK, is one of two introduced butterflies that now occur widely in North America, the other being the Small Cabbage White (*Pieris rapae*). It is a small, fast-flying butterfly found in grassy areas.

Females produce pale yellow eggs that are uniquely shaped, pill-like, and laid in strings on grass. They oviposit on many different kinds of grasses, although they prefer tall species; in Europe, cocksfoot grass (*Dactylis glomerata*) is a favorite. Unlike the eggs of most other skippers, European Skipperling eggs overwinter and hatch in spring. Curiously, the dormant embryonic caterpillar can be seen through the overwintering eggshell. The caterpillars are green and construct nests on their grass host plants. They pull a single blade of grass into a neat tube and then pin it together with tidy silk cross-ties. The caterpillars remain hidden to the world within these tubes by day and emerge at night to feed. Pupation also occurs within the tubes.

A SUCCESSFUL COLONIST

Introduced to Ontario, Canada, in 1910, this species is now common in the eastern USA and some parts of the Pacific Northwest. Sometimes hyperabundant and outnumbering all other butterflies, the European Skipperling is still expanding its range in North America. Ironically, perhaps, it is declining in abundance in the UK. It also occurs widely through Europe, Asia, and northern Africa. The species is common in agricultural fields, where it can occasionally be a pest of timothy grass (*Phleum pratense*). As dusk approaches, European Skipperlings move to sheltered sunlit areas of tall grasses, where they sometimes roost in the hundreds, often with multiple individuals sharing a single grass head.

European Skipperlings are exceptional colonists. Their spread is facilitated by their habit of feeding on widespread, common host plant grasses and their high mobility, with populations advancing up to 19 miles (30 km) annually in the USA. Another way in which populations of this butterfly spread is as dormant overwintering eggs on grasses dried for hay that are then transported long distances.

Caterpillar nest

Caterpillars of European Skipperlings construct nests by using silk from their mouthparts to tie together a blade of grass.

Blade of grass

Silk strand

→ The European Skipperling is a European export to North America where it can now be found in meadows and pastures in the northern USA and southern Canada.

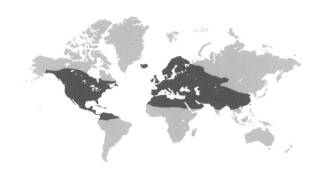

Red Admiral

Also known as Red Admirable

SCIENTIFIC NAME	*Vanessa atalanta* (Linnaeus, 1758)
FAMILY	Nymphalidae
NOTABLE FEATURES	Coal-black with scarlet bands
WINGSPAN	2–2½ in (50–65 mm)
HABITAT	Open habitat including parks, gardens, grassland, woodlands, and mountains

The Red Admiral is also known as the Red Admirable, which is perhaps a more fitting name since the admiral butterflies belong to the genus *Limenitis*, and the species *V. atalanta* is in the genus *Vanessa*. This charismatic and striking butterfly is, indeed, one to be admired.

The caterpillar of the Red Admiral leads a solitary and hidden life within a drooping leaf tent that it constructs on a stinging nettle (*Urtica dioica*) host plant. Feeding occurs within or outside the shelter at any time of night or day. Red Admiral caterpillars rarely wander from the host plant to pupate, and the pre-pupal caterpillars will sometimes make a leaf umbrella to shelter the pupa. In the UK, the species also uses hop (*Humulus lupulus*) as a host plant, although this is not the case in the western USA, where hop plants are often ignored. Females lay their ribbed green eggs singly, usually on the lower surfaces of nettle leaves. Caterpillars take about three weeks to develop and pupate.

PUGNACIOUS AND TERRITORIAL

With solid red, scarlet, or orange bands on a coal-black background, this butterfly is exceptionally striking. It is also very tough, able to live many weeks and even sometimes months as an adult. It may overwinter as an adult or migrate south.

The Red Admiral is widespread, occurring in North America, Europe, and Asia in a wide range of habitats, including gardens, parks, forests, meadows, orchards, and mountains. It is a strong flier and a lover of flowers, particularly butterfly bushes (*Buddleja*) in home gardens.

The male Red Admiral is pugnacious and territorial, perching on hilltops, bushes, or the ground, particularly late in the afternoon, from where it flies out to intercept intruders. Adults visit many kinds of flowers, and both sexes are also partial to feeding on rotting fruit, flows of tree sap, dung, and carrion. Feeding on fallen fruit is common in the fall, and may help the butterflies survive the winter or fuel migration. Red Admirals usually migrate northward in spring and south during fall, although some successful overwintering occurs in the UK and northwest North America, particularly in mild winters.

→   The Red Admiral has fabulous contrasting coal-black and scarlet-orange patterning. Alderman is yet another traditional English name for this butterfly, referring to the red and black clothes worn by aldermen in the past.

BUTTERFLY
BEHAVIOR

The flight of the butterfly

The flight of butterflies is a wondrous and amazing thing, about which scientists still have much to learn. Butterflies are able to fly just hours after emerging from their pupa, as soon as their wings have dried. Unlike birds, their parents do not teach them this skill and they must rely on instinct alone.

BUTTERFLY WINGS ADJUST BODY TEMPERATURE

Being cold-blooded, butterflies can fly only when the temperature is above a certain threshold, usually around 57°F (14°C). Nerve cells in a butterfly's wings can detect light and heat. They use the sun to regulate their body temperature, which is one reason they bask in its rays. In temperate areas butterflies bask to raise their temperature, while in the tropics where they can overheat, their wings can radiate heat to cool them off. Overheated butterflies can also seek shade and rest.

Basking enables butterflies in temperate regions to substantially raise their body temperature above ambient conditions, so they can fly when shade temperatures are cool as long as the sun is shining. Some species can warm up faster by shivering, which generates internal heat.

Temperate butterflies have also evolved other means of warming themselves up, including having dark, hairy bodies. The dark color absorbs heat, which is then trapped by the hairs. With their large surface area, the wings warm up relatively quickly and transfer heat to the body through the circulation of hemolymph.

← A Redspot Sawtooth (*Prioneris philonome*) takes flight on the Indonesian island of Sumatra.

↗ A Margined White (*Pieris marginalis*) dries its wings soon after eclosion in readiness for the first flight.

→ The dark, hairy body of the Blue Copper (*Tharsalea heteronea*) enhances the warming obtained by reflectance basking in sunshine.

Butterflies bask in different ways. Dorsal baskers, such as some Nymphalidae—including tortoiseshells, the Peacock (*Aglais io*), and sailers (*Neptis* spp.)—bask on flat ground with their wings open, often on a warm path. This traps warm air beneath the wings and body. Lateral baskers, such as many browns (Nymphalidae) and jezebels (*Delias* spp.), bask with their wings shut, but lean slightly to expose the undersides to the sun. Reflectance baskers, such as some whites (Pieridae), hold their wings at an angle to reflect the sun's warmth to the body.

In the middle of summer, butterflies may have the opposite problem: that of overheating. Once shade temperatures exceed 86–95°F (30–35°C), many butterflies seek shade, stop flying, and close their wings.

↑ The West Coast Lady (*Vanessa annabella*) is a dorsal basker, fond of basking with wings outspread on bare, warm ground, catching the sun's rays and trapping the warmth from the ground.

↗ The Western Green Hairstreak (*Callophrys dumetorum*) is a lateral basker, absorbing warmth through the undersides of its wings.

↗↗ Most blues (Lycaenidae) are reflectance baskers, directing the sun's rays and warmth to the body.

STRONG WINGS

Being able to fly gives butterflies a great advantage over wingless insects. In a second, they can be up and away from danger or on their way to finding nectar. Most people think that butterfly wings are extremely fragile, but the truth is that they are stronger and more resilient than one would expect—they have to be, to endure the physical stresses of flight, which may include journeys of hundreds of miles in migratory species. Today, with the aid of slow-motion cameras, scientists better understand the physics of butterfly flight.

Although a butterfly's wings appear rigid, in reality they are quite flexible. In flight, a butterfly rolls and twists its wings to create lift, thrust, and exceptional maneuverability. Although all butterflies share the same basic wing design and the physics of flight are universal, those from different families show a surprising range of flight patterns.

Small-winged skippers have fast, short, darting flights. Large brushfoot and swallowtail butterflies can fly rapidly by flapping their broad wings and soaring. Brown butterflies such as Ringlets (*Aphantopus hyperantus*) and Meadow Browns (*Maniola jurtina*) have a jerky type of undulating flight, usually fairly low to the ground and often within tall grass. Some butterflies use gliding as a major component of their flight. The sailers (*Neptis* spp.) glide above the vegetation with their wings open flat, rarely closing them to flap. On the other hand, many Satyrinae butterflies, including the genus *Ypthima*, somehow manage to fly with their wings closed above their heads most of the time.

Adult feeding habits

Flying, seeking mates, and producing offspring are all high-energy activities, so butterflies must feed. A newly eclosed butterfly does have some stored nutrients, carried over from the caterpillar stage, but these are usually expended within a day or two.

A SIMPLE DIET

The diet of a butterfly is simple: sugar (with perhaps a dash of amino acids), water, and salts. Flowers provide all of these requirements, and virtually all butterflies can derive the nourishment they need from a wide range of different flowering plants.

Nectar is a sugar-rich liquid contained in nectaries within flowers. Butterflies access the nectar with their proboscis, a hollow tube they use to probe nectaries and absorb nectar. It used to be thought that the proboscis functioned like a straw, sucking up nectar, but recent research has shown that it acts more like a sponge, absorbing the fluid.

INTOXICATING SUGAR SOURCES

Being avid flower feeders, butterflies get most of their sugar requirements from flowers. However, some species find alternative sugar sources, including sap flows from trees and honeydew, a sweet, sap-like juice excreted by aphids, mealybugs, and scale insects. Caterpillars of *Allotinus* and *Miletus* eat aphids. Adults of these species do not visit flowers, but imbibe honeydew of aphids, which they ate when younger. In the western USA, California Tortoiseshells (*Nymphalis californica*) are often seen on non-flowering conifer trees in late summer, probing needles and branches for honeydew from the aphids that live there. Other brushfoots, including Red Admirals (*Vanessa atalanta*), can be found in the fall feeding on overripe fruits that have fallen to

← The proboscis is a coiled, hollow tube that butterflies use to imbibe nectar and other nutrients. It absorbs like a sponge rather than as a suction device.

↗ A male brownie (*Miletus* sp.), uncurls its unusually short proboscis to feed on the honeydew of sap-sucking insects that are tended by *Dolichoderus* sp. ants.

the ground. These sugars have often fermented, and butterflies that drink to excess can become intoxicated. Such butterflies apparently lose their inhibitions and do not fly away when approached or if danger threatens.

FATS FOR ENERGY AND A LONG LIFE

Most butterflies use the sugars they obtain from flowers directly to maintain life and function, particularly flight and maturation of reproductive organs. In late summer and early fall, some species—including the North American Monarch (*Danaus plexippus*)—convert nectar sugars to lipids (fat), which they then use to fuel their famous migrations across hundreds or thousands of miles to overwintering sites in California and Mexico. Monarchs also store lipids to enable them to survive through the winter months in relative dormancy. Other species that overwinter as adults within their habitats, such as Mourning Cloaks and anglewings (Nymphalidae), also convert sugars to lipids to aid their survival.

PUDDLING PARTIES

Sometimes butterflies are seen gathered on damp soil or sand, seemingly so engrossed in feeding that you can get quite close to them. These gatherings are common wherever and whenever the weather is hot and dry. The most common participants in these so-called puddlings are swallowtails, whites, blues, and brushfoot butterflies.

Aside from obtaining moisture, puddling butterflies are also absorbing salts that they need for flight and metabolic functions. Salts are rare in flowers and are not accumulated by caterpillars. Curiously, nearly all puddling butterflies are males, but this is likely related to their need to provide salt and minerals to females as a nuptial gift at mating. Males with extra stored sodium are more attractive mates for females than those without the mineral, and have increased chances of offspring survival.

↑   Butterflies puddling in Brunei. Bornean Sawtooth (*Prioneris cornelia*), grass yellows (*Eurema* spp.), and Orange Gull (*Cepora iudith*) butterflies imbibe minerals from a damp area on the rainforest floor at Temburong National Park, Brunei.

→   Swallowtail butterflies like this Lime Swallowtail (*Papilio demoleus*) have long legs and a long proboscis, which allow them to exploit many different kinds of flowers.

FLOWER PREFERENCES

Different butterflies have different flower preferences. Some species clearly prefer flowers of a particular color, but many of the preferences are determined by flower structure and proboscis length. Butterflies with a relatively short proboscis can only gain nectar from flat, open flowers with shallow nectaries, such as daisies. Butterflies with a longer proboscis feed from tubular flowers with more deeply embedded nectaries, such as butterfly bush and thistles.

← A group of Great Blackveins (*Aporia agathon*) feed on dung in Nepal.

OTHER SOURCES OF MINERAL SALTS

Lepidopterists discovered long ago through personal experience that urine-moistened sand or soil is highly attractive to thirsty butterflies. Urine is a good source of sodium and "wild urine" is available from some online commercial sources as a butterfly attractant.

Another source of salt that butterflies will sometimes drink is human perspiration. Some butterflies will alight on a sweaty person and then probe their salty skin with their proboscis. Even sweaty clothes, boots, and hats worn on summer hikes can attract salt-seeking butterflies.

Some butterflies find salt in strange places. Crocodile tears may be insincere, but they are a good source of salt for some tropical butterflies, as are turtle tears. Another bizarre source of sodium that butterflies will take advantage of is blood—observers have reported that clothing soiled with fresh blood drawn by leeches has attracted feeding skipper and gossamer wing butterflies in India.

Salt and minerals can be obtained from another slightly off-putting source: animal dung. Clusters of butterflies feeding on animal dung are a common sight in tropical and subtropical areas, and in hot summers in more temperate areas. Another similarly dubious source of salts and minerals that butterflies sometimes use is carrion. Even insect carrion splattered over the front end of a car may attract feeding butterflies.

AN ACQUIRED TASTE

The most celebrated case of a butterfly specializing on unsavory foods is the Purple Emperor (*Apatura iris*; see pages 90–91). A few decades ago this species was rarely seen in the UK, but its numbers are now gradually increasing. Every summer, butterfly fans try to lure the Purple Emperors down to the ground in the hope of getting a good photograph, each using their own foul concoction of evil-smelling rotting fish, rancid cheese, and stinking diapers.

Resting and roosting

Like all animals, butterflies need to rest occasionally, particularly overnight. While most moths are nocturnal, some are able to fly by day, but it is rare for butterflies to fly by night. The exceptions are usually migrating butterflies such as Painted Ladies (*Vanessa cardui*) flying at high altitudes, which may continue flying or gliding during the night if temperatures aloft are warm enough.

← A clearwing butterfly in South America roosts for the evening.

↗ A Boisduval's Blue roosting in a head-downward position, typical of many blues in the family Lycaenidae.

NIGHTTIME IS FOR SLEEPING

At night, butterflies rest and roost. From an hour
to just minutes before sunset, they seek a nocturnal
roost. This may be in a bush or a tree, or simply near
the top of a grass stem or even on a flower—different
species have different preferred roosting locations. The
roosting preferences of many butterflies are unknown
and some may use a variety. The Red Admiral (*Vanessa
atalanta*) has been observed selecting a branch of a
conifer tree and then roosting in among its needles
for the night.

The species that choose flowers or grasses to roost
in—often blues (Lycaenidae)—are the most visible
because they sometimes roost gregariously. They make

a stunning sight as they sit, usually head-down, on grass
stems. Some species will spend the last hour or so of
sunlight basking with wings open at the roosting site,
and this is a good time to take photographs. They do
this again in the early morning, although they will shift
positions at this time to catch the new day's first rays
of sunshine. Many species do not become fully active
before 9 or 10 a.m. depending on the temperature
and season.

If the day is overcast and cool, butterflies remain
in their overnight roosts until temperatures increase or
the sun comes out. Similarly, if a sunny day turns cooler
and overcast, butterflies will find a suitable spot to rest.

FINDING A SAFE WINTER ROOST

Butterflies that overwinter as adults need to find
and choose a location that affords protection from
the elements, as well as from hungry winter predators
looking for a morsel of fat and protein. Overwintering
butterflies are full of fat stores, which they accumulate
during fall to help them survive a winter without
feeding. Many of the butterflies that overwinter as
adults are brushfoots, and these usually have plain dark
ventral hind-wing surfaces that enable them to blend
in with dark backgrounds in their winter roost.
Tortoiseshell butterflies and Peacocks (*Aglais io*) in
Europe often find a winter roosting spot in an open
building such as a shed or a stable, and multiple

butterflies are sometimes found roosting in a single
structure. Mourning Cloaks (*Nymphalis antiopa*) and
anglewings in North America find roosting sites in
tree holes or in hollow logs. Compton Tortoiseshells
(*Nymphalis l-album*) in Canada are sometimes found
roosting in piles of cut wood stored for winter use.

MAGNIFICENT MONARCH MEGA-ROOSTS

The prize for the most spectacular winter roosts
belongs to the Monarch (*Danaus plexippus*). The
incredible and enormous roosts of these butterflies
in a small area of central Mexico are well celebrated
in nature documentaries, books, and even movies.
The majority of the eastern North American
Monarch population—numbering in their millions—
roosts and overwinters in these small, high-elevation
locations. The population is estimated by counting
the number of hectares occupied by the roosting
butterflies, which completely cover every inch of
every fir tree, making branches sag from the
weight of so many butterflies.

Smaller but still spectacular roosts of
hundreds or thousands of overwintering
Monarchs occur at many locations in
coastal California, where some are
popular tourist attractions. In the
Sydney area of New South Wales,
Australia, the butterflies adorn
Eucalyptus and *Melaleuca* trees at
a number of sites year after year.

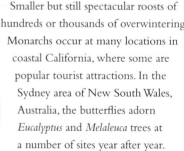

← The somber side of a Mourning
Cloak butterfly wing allows it to
disappear into darkness when
overwintering in a tree hole or shed.

→ Monarch butterflies aggregate
in hundreds and thousands at
overwintering sites along the
California coast.

Finding a mate

Finding a mate is the most important task of a male butterfly's life. How it does this varies considerably between and within butterfly families. Broadly, there are two main strategies: the "seek-and-find" approach and the "sit-and-wait" approach.

THE SEEK-AND-FIND APPROACH

Males that actively seek a mate will relentlessly fly along and patrol a defined and often linear route, such as a watercourse or a trail. Orange-tips (*Anthocharis* spp.), swallowtails (Papilionidae), and whites (Pieridae) often patrol the edges of woods and along ditches, hedgerows, and paths. These areas may have a high likelihood of being used by females for feeding or egg-laying, or they may be areas where females are eclosing from their pupae. Some butterflies, such as large brushfoots, use a long patrol route, while others, such as small gossamer wing butterflies (Lycaenidae), have a fairly short patrol route.

Butterflies that use a patrolling strategy to find females include the Monarch (*Danaus plexippus*), whites, and sulphurs, as well as some swallowtails. Male butterflies that patrol often have larger eyes than females of the same species. Once a patrolling male has detected a suspected female, he must take a closer look and check to smell if she has the correct pheromones.

Often a patrolling male butterfly will be seen chasing other similarly colored or similarly sized species, or sometimes other insects such as dragonflies and even small birds. Patrolling males sometimes encounter one another, when there may be a brief skirmish before they resume their patrols. Because of their incessant flight activity, patrolling butterflies are conspicuous.

Thus, the butterflies we see flying up and down a lane or through our yards are likely to be males. Females spend far more time resting, basking, and feeding than males, and are consequently less conspicuous. Patrolling males will briefly stop and sip nectar, but are soon on their way again—unlike females, which may spend many minutes feeding from a patch of flowers.

THE SIT-AND-WAIT APPROACH

Butterflies that take a sit-and-wait approach to finding a mate are usually quite territorial and will defend their perch from rival males and other flying insects. Perchers, by their nature, are less conspicuous than patrollers. Sit-and-wait males often establish their territories in specific areas, such as on a sunlit branch in a sheltered spot or on a rock in a stream.

→　Patrolling male butterflies, such as swallowtails (Papilionidae), chase females when they find them. A fit male and receptive female will mate after a short chase.

In the western USA, male Juba Skippers (*Hesperia juba*) often use rocks or stones in dried-up riverbeds as perches, darting out at any intruder within a space of a few feet. The best rock is much sought after, and the same male will use it day after day. Fights with interlopers are frequent, with the skippers flying high into the air before breaking off and the percher then resuming his sit-and-wait position.

Sit-and-wait males invariably perch in the sunshine and thermoregulate by adopting different positions and wing angles. They presumably do this to optimize their body temperatures so that they can successfully chase any females they encounter.

Are butterflies that defend a perch territorial? This is largely a semantic issue, but a percher is defending space for a mating rendezvous. As such, he is defending a resource and so this could be considered a territorial activity. Males chasing off interlopers is more investigative behavior rather than aggressive behavior. If the intruder is another male or a different species, the resident male returns. If it is a female of the same species, the male will pursue her, usually away from the perch site. Perchers may become worn and damaged during their short but frenetic lives. Male brownies (*Miletus* spp.) prefer a leaf in a sunfleck within the tropical forest where they live. As the sun moves across the sky, he moves to a new sun-dappled leaf and returns to the same spot day after day. A rival male may challenge him for a prime location by engaging in spiral, tandem flights.

It is possible that virgin females purposely fly into male-guarded territories so that they can be mated. Similarly, it is likely that mated females purposely avoid male-guarded territories. Unnecessary mating interrupts valuable egg-laying time and many female butterflies actively avoid multiple matings if they can.

↑ A Striped Albatross (*Appias olferna*) perches on an leaf.

← A male Viceroy (*Limenitis archippus*) butterfly perching on the top of a bush ready to fly out and intercept passing butterflies and other large insects, in the hope of finding a female.

Some butterflies combine both mate-finding strategies, spending some time sitting and waiting and other times patrolling. Individuals of species such as the European Speckled Wood (*Parage aegeria*) may choose to use one strategy to a greater or lesser extent, according to location or weather. Lorquin's Admiral (*Limenitis lorquini*) in the western USA will perch for many minutes, then fly off down a trail, turn around, and fly back to the same perch. He may encounter a female during one of these forays or dart at a passing female as he perches.

LEKKING

The males of some butterflies have yet another strategy, congregating with other males around certain landscape features to display for visiting females. This behavior is known as lekking, and is also seen in other animal species. Butterfly leks may form at unremarkable features such as patches of grass or certain trees, but to butterflies they are special places for sexual encounters.

In Europe, males of the Small Heath (*Coenonympha pamphilus*) gather in grassland, in places which may be slightly warmer than surrounding grassy areas and have a nearby landmark such as a tree or bush. Female Small Heaths search out these leks and fly conspicuously over them, soon attracting a mate in the process.

HILL-TOPPING

Another lekking strategy, used by males of many butterfly species across a number of families, is to find the highest point of ground in an area. In mountainous habitats, leks are formed on the loftiest, most exposed rocky summit. At lower elevations, the highest point of ground may be a gentle rise or even a man-made pile of gravel or debris. This behavior is commonly known as hill-topping. Often, the hill-topping site will be favored by more than one species, so it can become a very busy and colorful spectacle, with many butterflies swirling around a small, elevated piece of ground.

Swallowtail, white, skipper, and brushfoot males routinely hill-top. In most cases individual males defend large perching areas for two to three hours a day on the most prominent points on a rocky ridge. They do this by rapidly chasing off intruders, including in upward spiral flights, and the same male usually returns to defend the same site day after day. When a virgin female enters the zone, the male will immediately mate with her, usually without any courtship or extended chases.

↑ Female Small Heath butterflies, seeking attention, fly conspicuously over groups of males.

← Male Chalkhill Blue (*Lysandra coridon*) butterflies swirl around a female in a lekking and courtship display that will lead to one male becoming a successful partner for the female.

Butterfly reproduction

Courtship and mating are the highest priorities for butterflies once they emerge from the pupa. As with all species, passing their genes on to the next generation is their purpose on Earth. Butterfly lives are short, and unless reproduction is delayed to another season, it must take place within a short time frame.

FINDING A PARTNER

Male butterflies usually eclose before female
butterflies, ensuring that they are already active
and searching by the time females emerge. They
will sometimes even hang around the pupae
of females waiting for them to eclose. This is the case
with the Common Imperial Blue (*Jalmenus evagoras*;
see pages 246–247). As soon as a female begins to
eclose, fighting ensues, and the largest male usually
wins the right to mate with her. Most females are
mated within the first day or two of adult life.

Once a male and female meet, they both need
to be assured that their chosen mate is of the right
species and is fit and virile enough to produce an
optimal number of offspring. Butterflies check mate
suitability using scent, size, and appearance. Stamina
is also needed to complete and succeed in energetic
courtship flights. A female in the first days of adult
life will usually be a willing partner, as long as the
male ticks all the right boxes.

Males of many butterfly species use pheromones
to attract a female. It is presumed that each species
uses a different compound or pheromone mixture.
Many butterfly courtships take place quickly, but
some are elaborate, with prolonged aerial pursuits.
Males can dispense pheromones in different ways,
including via specialized scales on the wings, eversible
"hair pencils," or abdominal pheromone glands.

↑ A male Snowberry Checkerspot
(*Euphydryas colon*) attempting to get
the attention of a larger female with
a view to mating.

← A mating pair of Coronis
Fritillaries (*Argynnis coronis*). Males
of this species eclose a few days
earlier than females, and most females
are mated within 24 hours of eclosion.

REPRODUCTIVE STRATEGIES

The mating process can take anywhere from just a few hours to an entire night, so butterfly couples usually seek a quiet spot. The male passes the female a spermatophore, a small enclosed packet of sperm and nutrients. Producing the spermatophore consumes a significant amount of the male's resources. The transferred nutrients include sugars, salts, and amino acids that maximize egg production.

Studies on some species indicate that males may "linger" in copulation, even though transfer of the spermatophore may take only ten minutes. This leisurely mating strategy presumably reduces the time available for further mating. If disturbed, a mating couple can fly with one sex carrying the other, which hangs loose with wings tightly closed.

In nymphalid butterflies, the female is the flying partner, while in swallowtails and whites, it is the male. Unreceptive females usually try to avoid males, but females of butterflies in the family Pieridae display a rejection posture. In this, she raises her abdomen perpendicularly so that the male cannot make contact with his genitals. When pursued by males, some females will fly upward and then fall like a stone, a maneuver that often loses the male.

↑ Some butterflies may linger in copulation even though sperm transfer takes place quickly. It is thought males do this to reduce the chances of another male mating with the female.

MATING MONARCHS

A few butterflies, including the Monarch (*Danaus plexippus*), rely on force to land a mate. A male Monarch will pursue a female in a frenetic aerial encounter, aiming to literally grab her in midair with his spiny legs so that the pair fall to the ground together. There, he will forcibly copulate with the subdued female. Such is the male Monarch's drive and determination to mate that he will sometimes grab another male in midair, bring him to the ground, and attempt to copulate with him. Male Monarchs will even occasionally attack and try to copulate with other butterfly species, damaging them in the process. Celebrated naturalist Miriam Rothschild (1908–2005) famously called the male Monarch "nature's prime example of a male chauvinistic pig."

Not surprisingly, female Monarchs spend much of their lives trying to avoid interactions with males. They do this by visiting flowers early and late in the day when males are resting. They also spend a lot of time flying under the male radar by staying close to the ground among host plants and other shrubbery. To achieve optimal lifetime fecundity, a female Monarch needs to mate only once or twice, so excessive copulation reduces the time available for egg-laying, reducing lifetime egg production. In contrast, males are capable of mating every day or two.

Mated females invariably employ behavioral, chemical, and sometimes physical strategies to prevent multiple matings. In many cases, males will assist with this because it is in their best interest to father the progeny and not have their spermatophore superseded by that of another male. For example, males of around 1 percent of butterflies—mostly swallowtails and brushfoots—secrete a so-called "chastity belt," or sphragis, after mating, preventing other males from mating with that female and ensuring their paternity. Males of Small Cabbage Whites (*Pieris rapae*) include an anti-aphrodisiac pheromone with their spermatophore during mating. This decreases the attractiveness of the female to subsequent males, although it sometimes also has the unfortunate side effect of attracting small wasps that parasitize the eggs when laid.

↑ A female Mountain Parnassian (*Parnassius smintheus*) showing the white "chastity belt" (sphragis) secreted and placed by a male to ensure that his sperm will fertilize her eggs.

HOW BUTTERFLIES MATE

Butterflies mate by bringing the tips of their abdomens together, as seen here with this pair of Zebra Longwings (*Heliconius charithonia*). The male of this species has a pair of claspers that he uses to hold onto the female. The male then transfers to the female a capsule known as a spermatophore, which contains sperm. The female stores the spermatophore in a storage organ called a bursa copulatrix. Sperm then move to another vessel organ called a spermatheca. When eggs pass down the oviduct, sperm from the spermatheca fertilize them.

Butterfly mobility

Just like people, butterflies vary in their propensity to travel. Some species remain in the same location, while others are itinerants and range widely during their short lives. Still others make long-distance migrations.

"HOMEBODY" BUTTERFLIES

During their lifetime, some butterflies may fly only a few feet from the spot where they eclosed. Such species are habitat specialists, confining themselves to living within a specific site and using a specific host plant that may occur only in that habitat. As such, these species are vulnerable to changes in the habitat and usually have a greatly reduced ability to establish new colonies. Thus, these specialists may be readily extirpated, especially if a new habitat is not present within an easily accessible distance. Many blues, coppers, and hairstreaks (Lycaenidae) are habitat specialists, most likely because they often require specific ant mutualists in the same place as their host plants.

In reality, butterflies of discrete habitats usually rely on a mosaic of such habitats, with limited but important dispersal between the various patches.

← Ruddy Copper (*Tharsalea rubidus*) butterflies occupy small territories and live out their lives in very restricted habitats.

↗ The Firerim Tortoiseshell (*Aglais milberti*) ranges far and wide across the landscape in North America seeking nectar and host plant resources.

To date, the Xerces Blue (*Glaucopsyche xerces*) is the only butterfly known to have been extinguished in North America. This was an isolated population that had no linkage to a similar habitat, and died out because its San Francisco sand-dune habitat was consumed by urbanization.

ITINERANT BUTTERFLIES

Itinerant butterflies wander widely and freely, covering dozens of miles during their lives. These butterflies are typically generalists in terms of the habitat they require. Many are equally at home in wild meadows, gardens, and parks, as long as nectar is available. The host plants of itinerants are usually common across the landscape. Good examples of itinerant butterflies are brushfoots—including tortoiseshells and admirals, which feed as caterpillars on widespread nettles; and swallowtails such as the Lime Swallowtail (*Papilio demoleus*), which feeds on the leaves of citrus trees. Itinerant butterflies are often powerful fliers, which not only helps with their ranging dispersal, but also with avoiding the greater number of predators that they are more likely to encounter. Habitat generalists can effectively and rapidly exploit new breeding patches that may be many miles from where they originated.

APATURA IRIS

Purple Emperor

His Imperial Majesty

SCIENTIFIC NAME	*Apatura iris* (Linnaeus, 1758)
FAMILY	Nymphalidae
NOTABLE FEATURES	Males have a purple-blue, iridescent sheen
WINGSPAN	2¾–3¼ in (70–80 mm)
HABITAT	Open, deciduous woodlands

The Purple Emperor is one of Europe's most magnificent and charismatic butterflies, with iridescent purple wings. Often referred to as "His Imperial Majesty," it lives in open deciduous forests and is a butterfly of high summer and high trees.

This species is elusive, rarely descending from the treetops where it feeds on honeydew and tree sap. The male is breathtakingly beautiful, with an iridescent purple sheen that shows only at certain angles, a result of light being refracted from the wing scales. Males congregate in a lek, patrolling or perching to intercept any female that enters it. Occasionally, they come down to the forest floor, where they probe for salts and minerals in dung and wet soil.

To get a good view of a Purple Emperor, observers will set out all kinds of distasteful concoctions as bait to attract this species to the ground. Common baits include stinking shrimp paste, overripe cheese, rotting fish, roadkill, urine-soaked dog or fox feces, and even used diapers. Fish paste and fox feces seem to be the most successful. When a Purple Emperor is feeding on one of these baits, it becomes oblivious to its surroundings and is easily approached and photographed.

WILLOW TO TREETOPS TO WILLOW

Males patrol stands of willows (*Salix* spp.), the caterpillar host plants, looking for newly eclosed females. A virgin female will lead a pursuing male to the treetops, where mating takes place. Mated females visit patches of willows in the afternoon, often disappearing within the thicket to lay single eggs on leaves.

The caterpillars of Purple Emperors are rarely seen, hidden away in willow thickets. They are exquisitely camouflaged, as is the pupa, which mimics a willow leaf. The light-seeking adults usually emerge in mid- to late June.

→ The Purple Emperor is a showstopper. Entire books have been devoted to this most royal and charismatic European butterfly, despite its nauseating feeding habits.

ARGYNNIS PAPHIA

Silver-washed Fritillary

Flashy denizen of the forests

SCIENTIFIC NAME	*Argynnis paphia* (Linnaeus, 1758)
FAMILY	Nymphalidae
NOTABLE FEATURES	Orange and black with silver-streaked green hind wings
WINGSPAN	2¼–2¾ in (54–70 mm)
HABITAT	Open deciduous forest

The Silver-washed Fritillary is a spectacular large orange-and-black butterfly. It is found in summer swooping and gliding down trails, occasionally pausing to sip nectar from flowers. Silver streaks on the underside of this butterfly's wings give it its common name.

This butterfly has an elaborate courtship flight in which the male flies alongside the female, releasing pheromones. When mating takes place, the male transfers chemicals to the female along with the spermatophore to repel other males.

Silver-washed Fritillary caterpillars use violets as their food plants, feeding mostly by night on the forest floor.

The female does not lay her eggs on violets, trusting her offspring to find the plants themselves. However, she does give them a head start by making sure that violets are present on the forest floor in the area where she lays her eggs. She does this by walking and "tasting" (using receptors on her legs) for the presence of violets. She will then find a protected spot nearby to lay her eggs, often within the fissured bark of a tree. The caterpillars hatch in late summer but do not feed or leave the tree until after winter, remaining safe in the protected spot their mother chose. In the spring, the caterpillars climb down the tree in search of fresh new violet leaves to feed on.

A BUTTERFLY THAT IS WINNING

The Silver-washed Fritillary is one of a few European butterflies that is thriving in the face of habitat and climate change. Once a localized, rarely encountered species, it has now become a common sight in recent years in forests across England and Wales. The reasons for this are uncertain, but it is likely that the warming climate has at least something to do with it. It is important for this butterfly that forests do not become dense as it favors clearings and openings.

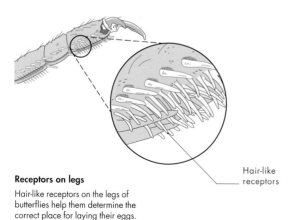

Receptors on legs

Hair-like receptors on the legs of butterflies help them determine the correct place for laying their eggs.

Hair-like receptors

→ The Silver-washed Fritillary is doing well in its European woodland domain, likely appreciating and thriving in warmer summers.

LIPHYRA BRASSOLIS

Moth Butterfly

A parasite of Asian weaver ants

SCIENTIFIC NAME	*Liphyra brassolis* (Westwood, 1864)
FAMILY	Lycaenidae
NOTABLE FEATURES	Large lycaenid with a stout body and orange wings
WINGSPAN	Male 2¾ in (71 mm); female 3 in (76 mm)
HABITAT	Open or slightly disturbed forests up to 650 ft (c. 200 m) in elevation

This widespread butterfly is rarely seen, perhaps because it spends most of its life as a caterpillar living inside arboreal ant nests. The adults seem to fly at twilight, but have been known to be attracted to ultraviolet lights at night, like many moths. The species ranges from the Himalayan foothills throughout all of subtropical and tropical Southeast Asia and Australasia, extending to the Solomon Islands.

Caterpillars of the Moth Butterfly feed on the eggs, larvae, and pupae of the Asian weaver ant (*Oecophylla smaragdina*) inside the ants' nest. Asian weaver ants make their nests by sewing living tree leaves into a ball using silk spun by their larvae. The ant queen and the colony's immature stages can be found inside. Females of the Moth Butterfly lay their eggs on or near a weaver ant nest, and the newly hatched caterpillar crawls inside to start feasting on the soft-bodied ant eggs, larvae, and pupae.

AN UNWELCOME GUEST

Many butterfly caterpillars that live with ants do so harmoniously. The caterpillars ply the ants with food rewards, employ chemicals that fool the ants, or manipulate the ants' brain chemistry via their secretions. Caterpillars of the Moth Butterfly, however, are attacked constantly by the ants.

Fortunately for them, their tough skin provides adequate defense. The caterpillar's setae, which are normally hair-like, are instead modified into flattened umbrellas. These overlapping plates, attached to the body via stalks, act as a flexible armor. The caterpillar stays inside the ant nest to pupate, and leaves its final caterpillar skin as a hardened, protective case around the pupa. When it ecloses, the ants start to attack the soft-winged butterfly, but it has one more trick up its sleeve. The entire body is covered with an extra set of long, hair-like scales, including the scales on the wings. These "deciduous" scales fall off at the slightest touch. As the butterfly makes its way to the best entrance, attacking ants get a mouth full of fluffy scales, which they stop to remove one by one. Once outside the nest, the butterfly drops to the ground and walks until it finds a stem or trunk it can crawl up to expand and dry its wings. Hours can elapse between eclosion and wing hardening, and the Moth Butterfly's wings remain unhardened for an unusually long time.

→ Newly eclosed Moth Butterflies are clothed in a temporary covering of fluffy scales that protect them from ants as they leave the nest where they pupated. This individual has found a safe tree branch where it can expand and dry its wings.

PAPILIO AEGEUS

Orchard Swallowtail

Citrus lovers

SCIENTIFIC NAME	*Papilio aegeus* (Donovan, 1805)
FAMILY	Papilionidae
NOTABLE FEATURES	Large black butterfly with red, white, and blue markings
WINGSPAN	4½–5½ in (120–140 mm)
HABITAT	Gardens, orchards, and parks

The Orchard Swallowtail is a handsome large black-and-white swallowtail found commonly in gardens in northern and eastern Australia, as well as in New Guinea and its surrounding islands. Males are mostly black, while females have larger areas of white and prominent, eye-catching red and blue spots.

Male Orchard Swallowtails patrol gardens and trails searching for females. After a brief spiraling courtship flight, mating occurs, during which, unusually, both partners keep their wings open. Females wander widely, looking for citrus plants and laying their spherical green eggs singly on the upper or lower surfaces of young leaves. Young caterpillars are black and white, and resemble fresh bird droppings. Older caterpillars are green and spiny, with distinct white and black areas, allowing them to blend in remarkably well with citrus foliage. To deter predators, they evert a forked, Y-shaped osmeterium gland from behind the head, which emits a pungent citrusy aroma. Although the caterpillars are not normally damaging to mature trees, large individuals can defoliate small saplings.

The wing patterns of this species are highly variable throughout its range. In areas where owl butterflies occur (*Taenaris* spp.) in New Guinea and several nearby islands, male and female Orchard Swallowtails mimic them. The wings are predominantly white or pale gray, and the bottom of the hind wings sport boxy splashes of color—the swallowtail's attempt at reproducing the large eyespots of the owls.

A TASTE OF THE TROPICS

These large, black butterflies always attract attention when they visit citrus and their favorite flowers in gardens, particularly in more southerly areas of Australia, since they seem so flashy and downright tropical. They hover above flowers like outsized hummingbirds while their long proboscis probes for nectar. They are strong but haphazard fliers, capable of rapid flight if disturbed. This is the swallowtail that most Australians know since it so often shares the outdoors with them.

→   The Orchard Swallowtail is one of eastern Australia's most common and showy butterflies, often living happily with people in suburbia.

PARNASSIUS SMINTHEUS
Mountain Parnassian
Companion of mountain hikers

SCIENTIFIC NAME	*Parnassius smintheus* (Doubleday, 1847)
FAMILY	Papilionidae
NOTABLE FEATURES	High-altitude white butterfly with red and black markings
WINGSPAN	2¼–2½ in (56–64 mm)
HABITAT	Mountain meadows, rocky ridges, and summits

The Mountain Parnassian can be found in the Rockies of western North America, flying effortlessly through alpine meadow habitats. Females are kept chaste by the first male they meet, which secretes a mating plug to stop her mating again.

Females lay their large, round, white eggs on objects near stonecrop plants (*Sedum* spp.). The embryonic caterpillar forms in a few weeks, but remains dormant in the egg for six to eight months before hatching when the snow melts.

The caterpillars feed on stonecrop plants and are black with yellow dots, well designed to absorb heat in the cool mountain climate. They resemble a distasteful millipede with black and yellow dots, which they are likely mimicking for protection. If you hike in the mountains, you may find Mountain Parnassian caterpillars sunbathing on the ground or on rocks. They do this to raise their body temperature and speed development.

REPRODUCTIVE STRATEGY
Males cruise the landscape looking for females, which they invariably find on the ground. Mating lasts for several hours, after which the male secretes a waxy pluglike structure, called a sphragis, and fits it to the female's lower abdomen. This prevents the female from mating again, ensuring that only his sperm will fertilize her eggs, thus preserving his genetic contribution to the next generation.

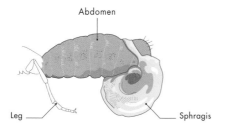

Abdomen

Leg

Sphragis

Sphragis in a Mountain Parnassian
The male Mountain Parnassian secretes a sphragis that he attaches to the female abdomen after mating to preserve his genetic legacy.

→ The Mountain Parnassian is a true mountain butterfly found at the highest elevations where its stonecrop host plants grow.

HABITATS & RESOURCES

Home is where the habitat is

A butterfly's habitat is the place where it normally lives. It may be small like a hillside, or large like a geographic region. Its habitat provides a butterfly with all the essentials for life, including food, shelter, host plants, mates, and optimal climate.

THE IMPORTANCE OF HABITAT

Every creature lives in a habitat. Some are specialists, never leaving the place where they were born. Others are generalists, spending their lives traveling from habitat to habitat, each of which may be different.

Butterflies occupy a wide variety of habitats, including grasslands, forests, urban areas, mountains, wetlands, deserts, and tropical forests, to name but a few. Some species—the generalists—can be found in many, if not all, of these habitats, while others— the specialists—are restricted to specific habitats.

A habitat must provide the resources a butterfly needs for survival. Consumables such as nectar and caterpillar host plants are obviously essential, but also needed are "utilities," including shelter, basking areas, courting areas, and places safe from predators.

SPECIALISTS AND GENERALISTS

Butterfly diversity in relation to habitat ranges from extreme specialists such as the rare Leona's Little Blue (*Philotiella leona*; see pages 146–147), which is restricted to just 15 square miles (39 sq km) of high desert habitat in Oregon, USA, to the common Painted Lady (*Vanessa cardui*), which can be found on all continents except

← A group of Large Striped Swordtail (*Graphium antheus*) males imbibe muddy water to extract sodium at a watering hole in Gorongosa National Park, Mozambique.

↗ Leona's Little Blue has one of the most restricted butterfly habitats in the world, confined to just 15 square miles (39 sq km) of high desert habitat in Oregon, USA.

→ The Common Cerulean (*Jamides celeno*) is at home in disturbed, open habitats in Southeast Asia.

Antarctica and in many habitats. However, most butterflies fall somewhere in between these extremes, inhabiting a subset of habitat types throughout a species' range.

In the UK, habitat generalist species are often referred to as wider countryside species and are usually described as widely distributed. However, habitat specialist species can also be widely distributed if their special habitats can be found across the landscape. Caterpillars of the Bog Copper (*Lycaena epixanthe*) feed on the leaves of cranberry plants (*Vaccinium* spp.), which grow in waterlogged, acidic bogs. In addition, adult butterflies of this species feed exclusively on nectar from cranberry blossoms. Although its host plant, nectar plant, and habitat requirements are strict, suitable bogs are found throughout the northeastern United States and southeastern Canada. Therefore, its range encompasses hundreds of thousands of square miles (or square kilometers) of land, even though its distribution within that area would be described as "patchy" because of the species' habitat specificity.

THE RISKS OF BEING SPECIAL

Many habitat specialist butterflies are restricted in
distribution when the habitat they need is also localized
and rare. The presence of host plants and favorable
climate are the two major factors determining the
suitability of a habitat to a specialist butterfly. Without
the right host plant, a butterfly population cannot
survive, and if the host plant is rare, so the butterfly
will be too.

This is the case for Leona's Little Blue, which
depends on the uncommon and localized spurry
buckwheat (*Eriogonum spergulinum*) for caterpillar
development. However, favorability of climate is likely

a cofactor, as the butterfly also needs a hot, sunny flight
period in summer. In reality, host plant presence and
climate favorability are intrinsically connected in most
specialist butterfly habitats.

Populations of habitat specialist butterflies are
predominantly "closed," which means most individuals
live and die in the same area. When they reach the
edge of their habitat, they will turn around and come
back. However, there will always be a few individuals
that will wander out of the habitat. These are
important because they are potential founders of new
populations and maintain genetic diversity. As we will
learn in Butterfly Populations, discrete populations

← The Common Eggfly (*Hypolimnas bolina*), like this female in Java, Indonesia, thrives in the disturbed habitats that now dominate much of tropical Asia and the South Pacific.

→ The large Two-tailed Tiger Swallowtail (*Papilio multicaudata*) is capable of powerful flight and wanders far across the landscape.

cannot exist in perpetuity. There must be some connection to other populations, and the individuals in a population that roam from their habitat are key to this linkage.

THE ADVENTURE OF BEING A GENERALIST

Habitat generalist butterflies have "open" populations, as they range far and wide across the landscape. When tortoiseshells, anglewings (Nymphalidae), and swallowtails (Papilionidae) encounter a barrier such as a forest or large body of water, they may fly over or around them. Wide-ranging butterflies will often use natural or man-made corridors such as roads, trails, field margins, rivers, and powerline rights-of-way to navigate across the landscape. These corridors are also valuable in offering opportunities for shelter and feeding.

Different habitats—for example, forests and alpine areas—have characteristic assemblages of butterfly species. This means that butterflies adapted to living on the tops of mountains would not do well in a warm, humid swamp—although the Painted Lady (*Vanessa cardui*), an exceptional generalist, might be seen in both habitats. In the following pages, we will explore the kinds of butterflies that can be found in different habitats and see how their lives are adapted and in tune with their chosen home.

Butterflies of deciduous forests

Deciduous forests in temperate regions are important butterfly habitats in summer, especially if they are relatively open, allowing the sunshine to percolate into glades. An understory of grasses and herbaceous plants is also important, to provide nectar and host plants.

COMPLEX HABITATS

Temperate deciduous forests can be great places to see butterflies, as long as the canopy is not dense and there are plenty of flowery meadows. This habitat type offers plenty of opportunities for shelter and protection for butterflies, both from the weather and natural enemies.

Forests can be complex habitats containing different kinds of microhabitats, including edges, open patches of grass, and shady areas. Many forest butterflies have declined in numbers because of a trend in forestry toward less management, making the habitat denser and shadier. Within forests, most butterflies are found along edges and in sunlit gaps, and if they are rich with flowers then they will also be rich with butterflies and other insects.

Butterflies from all families found in temperate regions are represented in deciduous forest faunas, particularly blues (Lycaenidae), brushfoots (Nymphalidae), and skippers (Hesperiidae). An estimated 20 percent of North American butterflies can be found in mixed deciduous forests. Among UK and European butterflies, the number is even higher—perhaps 50 percent or more.

← A pair of Silver-washed Fritillaries (*Argynnis paphia*; see pages 92–93) flying in copula through their forest habitat.

↗ Lorquin's Admirals (*Limenitis lorquini*) patrol grassy forest trails looking for mates.

↗↗ The Wood White (*Leptidea sinapis*) flies slowly but determinedly along wide and sunny forest paths in Europe and Asia.

BUTTERFLY DIVERSITY IN FORESTS

Characteristic butterflies of North American deciduous
forests include tiger swallowtails, blues, checkerspots, admirals,
browns, and skippers. A similarly diverse butterfly fauna is characteristic
of European forests. Most forest butterflies are found in open, grassy,
flowery areas, but they can also be seen flying along forest trails as they
seek partners or egg-laying sites.

A common and familiar forest butterfly in mainland Europe
and the UK is the Speckled Wood (*Pararge aegeria*; see pages
144–145). This is one of the few species that has done well in past
decades and is expanding its range. It relishes the shady conditions
and dappled sunlight found in most forests, but will also venture
into meadows, parks, and hedgerows. This is one species that may
have benefited from the shadier conditions created by reduced
forest management.

↑ Temperate forests are rich places for butterflies as long as they have plentiful meadows, trails, and clearings for flowers and host plants.

↗ Temperate grasslands with their bountiful sunshine and diverse flora are invariably full of butterflies during the summer months.

Butterflies of shrub-steppes and deserts

Hot, dry areas are not usually places associated with butterfly abundance and diversity. However, deserts and the natural arid grasslands known as shrub-steppes can sometimes be teeming with butterflies, particularly during spring or whenever soil moisture levels are greatest.

HOTSPOTS OF BUTTERFLY DIVERSITY

Surprisingly perhaps, deserts are productive habitats for butterflies, with many species adapted to the rigors of living in a dry and frequently desiccated environment. Similarly, in shrub-steppe habitats in North America and Europe, which are essentially low-rainfall natural grasslands, butterflies sometimes abound, particularly during spring.

Recent studies show that the deserts of the southwestern USA and Mexico are butterfly diversity hotspots, with high numbers of endemic species despite poor diversity of flowering plants. This might be explained by a limited number of host plant species serving multiple butterfly species. To live in a desert, a butterfly must be highly adapted to extreme heat and aridity. It must also be able to cope with unreliable host plants, both in terms of availability and quality.

← The opals, such as this Burnished Opal (*Chrysoritis chrysaor*), thrive in the dry fynbos of southern Africa.

HIGH DESERT SPECIALIST

Leona's Little Blue (*Philotiella leona*) lives in a restricted desert habitat in western North America, which means it has to survive a hot, dry summer and a bitterly cold winter. The species spends most of its life (ten months or more) as a pupa lying exposed on the sandy soil surface. These pupae are exposed to daily average soil temperatures that exceed 158°F (70°C) for at least two months, and then may experience temperatures well below 32°F (0°C) for five months or more. The butterflies emerge in mid-summer, and then have just a few weeks to lay their eggs so that the caterpillars have time to complete their development on the buds and flowers of their buckwheat host plants before the flowers wither.

→ Leona's Little Blue basking on a warm stone to raise its body temperature sufficiently to allow flight.

HEDGING THEIR BETS

Many butterflies that dwell in desert or shrub-steppe habitats use a bet-hedging strategy. This means, for example, that not all Silvery Blue (*Glaucopsyche lygdamus*) and Leona pupae produce butterflies the following year; instead, some remain dormant for multiple years. Montane and desert-shrub-steppe environments are often unpredictable, especially in terms of whether seasonal rainfall arrives on time—or even at all.

Many shrub-steppe butterflies reproduce in spring, when soil-moisture levels are highest and host plants are in optimal condition. Dry winters and springs reduce the quality and amounts of host plant foliage available, which may have an impact on the survival and development of shrub-steppe butterflies such as Sheridan's Green Hairstreak (*Callophrys sheridanii*) in the inland northwest of North America. Like Leona's Little Blue, Sheridan's Hairstreak overwinters as a pupa and some do not emerge in spring, instead spending a further year or more as a dormant pupa. This strategy, whereby a proportion of the population does not emerge in any particular spring, is designed to avoid the loss of that population should seasonal conditions prevent caterpillar development.

RIPARIAN RESOURCES AND REWARDS

Desert and shrub-steppe habitats usually incorporate natural areas
that follow the routes of watercourses or occur around wet areas
such as seeps and ponds. These riparian habitats offer additional
resources and rewards to arid-habitat butterflies in terms of nectar,
salts, minerals, host plants, and a preferred microclimate.
Consequently, higher butterfly populations may occur in these
zones and certain species specialize in this sub-habitat type, even
sometimes managing to produce more than one generation
because their host plants stay green longer.

In western North America, the Viceroy (*Limenitis archippus*),
an admiral butterfly that mimics the Monarch (*Danaus plexippus*)
in coloration and markings, is largely confined to riparian zones
within arid landscapes. The willow host plants needed by this
species occur only in riparian areas.

↑   The Viceroy is a North
American butterfly of riparian
habitats that utilizes willows
as caterpillar host plants.

↗   Early instar caterpillars of
checkerspots (*Chlosyne* spp.)
spend the hot summer of the
shrub-steppe congregated and
dormant at the base of their
host plants.

A CHANGING COLOR PALETTE

The shrub-steppe environment can be shades of brown or green depending on the season. During the short spring, green abounds as plants burst forth, along with the white, yellow, blue, and reds of flowers. Once summer arrives, the colors turn to orange, brown, and gray, and remain that way until the following spring. Some butterflies track this color change. For example, the green pupa of orange-tips (*Anthocharis* spp.) formed on green vegetation in mid-spring turns brown and then gray as plants wither and the seasons change.

First-instar caterpillars of the Desert Checkerspot (*Chlosyne acastus*) spend the late spring feeding on host plants and are greenish, but once the summer heat hits, they stop feeding, turn gray to black in color, and congregate at the base of their host plants, spending the next eight months in torpor.

The race to complete development of eggs, caterpillars, and pupae in desert and shrub-steppe environments sometimes results in undersized adults. This is because food runs out before the caterpillars are fully fed. Adaptation to this uncertainty over resources includes the earlier formation of pupae than would be the case if food were plentiful. These smaller pupae produce smaller adults, thus undersized butterflies are a feature of many desert species.

WINTER HOME OF PAINTED LADIES

Although the Painted Lady (*Vanessa cardui*) (right) is the most widespread butterfly in the world and occupies virtually all habitats, from urban gardens to mountaintops, populations in North America and Europe are restricted to desert-type habitats for part of the year.

Winter populations breed on host plants in desert fringes of sub-Saharan North Africa and arid lands in Mexico and the southwestern USA. Good seasonal rainfall in these deserts produces lush and abundant host plants during winter, and Painted Ladies breed prolifically, producing migrants in spring that fly north to occupy northern Europe and North America during the summer.

Butterflies of tropical forests

Butterfly diversity and abundance peak in tropical forests. Recognized as faunal and floral diversity hotspots, tropical forests are home to hundreds of butterfly species within limited geographic areas.

LARGE AND SPECTACULAR BUTTERFLIES

Tropical forests are home to a variety of large, colorful butterflies of many species. The spectacular blue morphos (Nymphalidae) and enormous birdwing (Papilionidae) butterflies (with a wingspan of about 12 in/30 cm) are among the most celebrated species of tropical rainforests, but there are hundreds of equally stunning butterflies to be found in this habitat. It is estimated that about 90 percent of the world's butterfly species are found in tropical forests, with abundance and diversity reaching a zenith in the tropical forests of South America. The species diversity of tropical Asia and Australasia are close behind, followed by the Afrotropics. Tropical mountains, such as the Andes of South America and the Central Highlands of New Guinea, boast the highest diversity per unit area.

→   The Blue Morpho (*Morpho helenor*) is an iconic species of the tropical rainforest. Spectacularly iridescent, these butterflies are found in South and Central America.

← Tropical brushfoots and other butterflies including this Malay Cruiser (*Vindula dejone*) in Sulawesi are commonly seen imbibing muddy water from puddles and streams to obtain sodium.

→ Many tropical satyrs and other brushfoots, such as this Common Evening Brown (*Melanitis leda*) in Sulawesi, prefer to feed on ripe or rotting fruit instead of flowers.

Tropical rainforests receive at least 79 in (200 cm) of precipitation per year and are situated within 20 degrees north or south of the equator, so they are always warm and wet. Aseasonal forests nearest the equator receive at least 4 in (10 cm) of rain every month, continually providing optimal growing conditions. However, some rainforests are punctuated by a dry season that may last up to half a year. A mature lowland tropical forest consists of several strata, from the ground to the tree canopy.

DIVERSITY ACCORDING TO HEIGHT AND LIGHT

Tropical forests host different assemblages of butterflies from ground level to the canopy, dictated by different levels of light and heat in each of the layers. Preferences for a specific vertical stratum might come from the host plant or from the butterfly itself. The types of butterflies living in gaps within the forest are invariably different from those in the dense forest, and the diversity here is greater. A large proportion of tropical adult butterflies use non-floral food sources, particularly fruit. For

example, some brushfoot butterflies are almost entirely fruit feeders as adults, especially those that live in the upper canopy.

THE SECRET OF LONGER LIFE

Feeding on a diversity of food sources has enabled most adult tropical butterflies to live longer lives than their temperate counterparts. Longwing butterflies (*Heliconius* spp.) uniquely feed on pollen, which releases amino acids into their proborces and extends their life span to as long as six months, allowing them to produce an increased number of eggs. Tropical butterflies can often continue breeding throughout the year, although some tropical forests—such as those in northeastern Australia—do have a dry season, which their inhabitants must endure. They do this either by migrating to more suitable habitats, or by shutting down reproduction and waiting for more favorable conditions.

↑   The Thick-Tipped Greta (*Greta morgane*) butterfly is a tropical species primarily found in the rainforests of Central and South America.

↗   The Zebra Longwing (*Heliconius charithonia*) is primarily a tropical butterfly found in South and Central America, but can sometimes be found in the far south of the USA.

Butterflies of grasslands and meadows

Second only to tropical forests in terms of butterfly diversity and abundance are flowery grasslands and meadows. Caterpillars of most skippers and browns eat grass, and these groups are particularly abundant.

A BUTTERFLY-RICH HABITAT

Unfertilized semi-natural meadows and savannahs
are a rich habitat for butterflies in temperate regions,
including Europe and North America. Some of these
grasslands have been used for centuries for livestock
grazing, and this has allowed a diverse flora of forbs
and other low-growing flowering plants to develop,
providing host plant and nectar resources for a
community of grassland butterflies. Many of these
species are in the brushfoot (Nymphalidae) and
gossamer wing (Lycaenidae) families.

In mainland Europe and the UK, calcareous
(chalk and limestone) grasslands, acid grasslands,
and wet grasslands all provide unique habitats for
butterflies. These chalklands are among the best places
to see butterflies, being home to Chalkhill Blues
(*Polyommatus coridon*), Marbled Whites (*Melanargia
galathea*), Ringlets (*Aphantopus hyperantus*), Duke of
Burgundies (*Hamearis lucina*; see pages 274–275),
skippers, tortoiseshells, Common Brimstones
(*Gonepteryx rhamni*), and a host of other species.

THE IMPORTANCE OF MICROHABITAT

Vegetation structure is important for determining the
butterfly species composition in grassland habitats.
Certain species, including some skippers, prefer
shorter turf as this provides greater warmth, while
others, including the Ringlet, prefer taller growth
and a cooler microclimate.

← The Common Brown (*Heteronympha
merope*) butterfly lives up to its name in the
savanna grasslands of southern Australia,
where sometimes populations can reach
high densities.

↗ The Chalkhill Blue (*Polyommatus
coridon*) is a common butterfly of chalklands
in Europe and the UK. At some sites,
thousands fly during the summer months.

→ The striking two-tone Marbled White
(*Melanargia galathea*) is a brushfoot
butterfly rather than a white (Pieridae),
and is common in unimproved grasslands
throughout most of Europe.

The Duke of Burgundy needs some bushes and scrub to provide a more humid environment and shelter. Warmer, south-facing slopes have different communities of butterflies than cooler, north-facing ones.

Chalk and limestone grasslands have been grazed for centuries, both by natural herbivores such as rabbits and by livestock. This activity is important to prevent plant succession, which would obliterate many of the low-growing host plants the butterflies that live here rely on.

ACID AND WET GRASSLANDS

Acid grasslands in Europe are formed on poor, thin soils, often overlaying sandstone or granite, and support a less diverse community of plants. Consequently, the diversity of butterfly species here is also less rich.

Bracken (*Pteridium aquilinum*) is frequently a dominant plant on acid grasslands, and the broken shade it provides allows the growth of violets, which are host plants of fritillary butterflies (Nymphalidae). Thus, acid grasslands are important habitats in the UK for some striking fritillaries such as the Dark Green Fritillary (*Argynnis aglaja*) and High Brown Fritillary (*Argynnis adippe*). Grazing is important in this habitat, too, to ensure the bracken does not become too dense, which limits or prevents violet growth.

Wet grasslands occur on heavy, poorly drained soils or alongside seepage areas, rivers, and streams. Grasses here are usually tall and lush, providing good habitat in Europe for grass-feeding species such as Meadow Browns (*Maniola jurtina*) and Ringlets. Conditions are cooler in this habitat, and some grazing is needed to provide areas where flowers can grow and bloom.

Wet grasslands are also important butterfly habitats in other parts of the world. In western North America, for example, they are a necessary component of riparian habitats along and near watercourses, supporting unique butterfly communities compared to the surrounding arid landscapes.

DIFFERENT COUNTRIES, DIFFERENT GRASSLANDS

Tallgrass prairie is an important but endangered butterfly habitat in central parts of North America. It supports some unique and declining species, including the Regal Fritillary (*Argynnis idalia*) and Dakota Skipper (*Hesperia dacotae*). Prescribed fire is an important management tool for North American prairies, but it needs to be carefully timed to have minimal impact on butterfly populations.

Savanna grasslands in Australia have a less diverse butterfly fauna than comparable European and North American grasslands, but are still home to a good number of species. The aptly named Common Brown (*Heteronympha merope*; see pages 200–201) is common in all kinds of grassland–forest mosaics and sometimes develops large populations.

← A grassland habitat incorporating at least some small bushes for shelter, is preferred by the Duke of Burgundy (*Hamearis lucina*).

↓ Most satyrs, like this Common Bush Brown (*Bicyclus safitza*) in Africa, feed on grasses and therefore make grasslands their home.

THE WRONG KIND OF GRASSLAND

One kind of grassland that is not a good butterfly habitat is most likely the one right outside your window. Yes, the all-too-common suburban lawn is one habitat that butterflies entirely eschew— unless, of course, you have a wildflower patch growing in it, in which case you may attract a passing butterfly to sip from the flowers' nectar. No Mow May, an initiative in parts of the USA and the UK that encourages those with lawns to leave them to grow wild for a month, throws an important lifeline to all urban spring pollinators, including butterflies.

Butterflies of mountain habitats

Temperature, precipitation, solar radiation, air density, and many other variables change as one climbs a mountain. The climatic changes experienced by ascending 500 feet (c. 150 m) in elevation is roughly equal to what you would experience by traveling 1 degree in latitude toward the nearest pole.

In temperate regions, this means that alpine summers are short and butterfly activity is crammed into a few short weeks. In the tropics, this means that alpine climates are similar to temperate conditions. Consequently, the closest relatives of some tropical alpine butterflies fly at much higher latitudes. For example, clouded yellows (*Colias* spp.) occur at high elevations in the South American Andes, and are among the most common butterflies throughout temperate North America and Eurasia.

ALPINE HABITATS

Surprisingly perhaps, mountains are often one of the best habitats for butterflies in terms of abundance and diversity. Spending a day in the mountains between June to August will often yield 50–60 species of butterflies in North America. So, it is no surprise to learn that more research has been done on alpine butterfly species than on those living in just about any other habitat.

Living in an extreme alpine habitat demands major adaptation. The summer here is always short (8–12 weeks in most places)

← Now believed to be geologically young—only around 5 million years old—the high peaks of New Guinea's central highlands are nevertheless among the most species-rich habitats in the world.

↗ The Hoary Anglewing (*Polygonia gracilis*) overwinters in high elevation forests and may live for more than ten months as an adult.

and much must be achieved during this busy period. There are flowers to visit, mates to find and court, and eggs to lay, all while trying to avoid equally frenetic natural enemies. A few hardy species, such as the Hoary Anglewing and Mourning Cloaks in North America, overwinter as adults and can be seen flying over the still-snowy landscape in spring, seeking the first nectar of the season.

However, the majority of mountain species overwinter as eggs or caterpillars, and there is no point for these species to "wake up" until the snow melts and new green shoots of their host plants appear. Depending on elevation, this may not happen until early summer, so the pressure is on for caterpillars to eat quickly and develop rapidly, lest they run out of time. Some species are not able to complete their development rapidly enough, so they overwinter for a second time. This means that adults of these species, which include the Astarte Fritillary (*Boloria astarte*) in northwestern North America, fly only every other year.

FINDING WARMTH IN THE MOUNTAINS

The life of a caterpillar depends on having sufficiently high temperatures to enable its development through four to six instars and into a pupa. Finding such warmth in the mountains can be challenging, however. Clouds often cloak high ranges, making ambient temperatures cool. Caterpillars struggle to eat and grow at temperatures below about 50°F (10°C), so they must take advantage of any warming sunshine they can soak up.

Dark colors absorb heat more readily than lighter colors, so most mountain-dwelling caterpillars are brown, gray, or black, and they bask and feed in direct sunlight. Caterpillars of the Mountain Parnassian (*Parnassius smintheus*; see pages 98–99) in western North America are a good example of this, being frequently found draped across small rocks and stones in early summer as they warm their mostly black bodies.

Even at the height of summer in the mountains when butterflies are flying, there are days when the sun is not shining and the temperature may be below the flight threshold, which in most alpine butterflies is around 54–55°F (12–13°C). However, with the sun high in the sky there is usually some radiant warmth percolating through the clouds, and many alpine butterflies are adapted to absorb this radiation to power their flight muscles.

← The Astarte Fritillary (*Boloria astarte*) is an arctic-alpine butterfly frequenting high ridges and tundra in upper latitudes of North America.

↗ The Pink-edged Sulphur (*Colias interior*) inhabits clearings and meadows in mid-high elevation forests in subarctic Canada and northern USA.

↗↗ The caterpillar of the Mountain Parnassian spends a lot of time basking in the sun to elevate its body temperature in a high-elevation habitat.

ALPINE ADAPTATIONS

Alpine butterflies are often more melanic (darker) than lower-elevation species. For example, sulphur butterflies (Pieridae) in the mountains have an abundance of black scales on their yellow wings; this allows the wings to absorb more heat, which is then directed to the body and flight muscles. Many alpine butterflies are also very hirsute. The hairs trap warm air, further aiding thermoregulation and flight. Melanic alpine butterflies fly more and feed more than lighter-colored individuals of the same species.

Mountain butterflies tend to emerge over an extended period. This may reflect the relative success of different caterpillars in finding warmth, which translates into varying developmental rates. Adults of many species survive burial under snow for a few days, a not-uncommon event in high-elevation habitats. This is unlikely to be seen in any lowland butterfly, most of which have trouble surviving a frosty night.

Butterflies of urban landscapes

Human-created urban landscapes have replaced natural habitats for some butterfly populations. However, there are growing opportunities for making this landscape friendlier to butterflies, and one in which they can live their lives successfully.

BUTTERFLIES IN TOWNS AND CITIES

Until relatively recently, city councils and municipal bodies had a policy of "tidying up" the "weeds" growing in parks, around sports fields, and in other urban green spaces. Thankfully, we have come a long way since those days. People and local governments are now aware and amenable to the idea of incorporating butterfly habitats within urban landscapes. Parks, gardens, golf courses, cemeteries, and other amenities are increasingly becoming better homes and habitats for butterflies, by allowing butterfly host plants (including some "weeds") to be part of their floral diversity.

Some butterflies are well adapted to becoming urban residents. Species that range widely and are habitat generalists are basically opportunists that can and will take advantage of host plant and nectar resources provided for them in urban areas.

→ Grass yellow butterflies (*Eurema* spp.) are common in urban gardens and parks throughout Asia.

(Inset) Even though its caterpillars feed on plant parasites, the Painted Jezebel (*Delias hyparete*) is nonetheless commonly seen flying through the streets and parks of even the most crowded tropical Asian metropolises.

↑ Orange Sulphurs (*Colias eurytheme*)
are frequent flower visitors in the urban
landscapes of North America.

In Europe, such butterflies include tortoiseshells,
Painted Lady (*Vanessa cardui*), Peacock (*Aglais io*), Red
Admiral (*Vanessa atalanta*), cabbage whites (*Pieris rapae*
and *P. brassicae*), Brimstone (*Gonepteryx rhamni*), and
Meadow Brown (*Maniola jurtina*). In North America,
tiger swallowtails, Monarch (*Danaus plexippus*), Painted
Lady, commas, and sulphurs are all common urban
butterflies. In Southeast Asia, these include Psyche
(*Leptosia nina*), Lime Swallowtail (*Papilio demoleus*),
Painted Jezebel (*Delias hyparete*), grass yellows (*Eurema*
spp.), and grass blues (*Zizula hylax* and others).

BUTTERFLY GARDENS

The famous British politician Winston Churchill
(1874–1965) was one of the first public figures to
cultivate a butterfly garden, which is now something
thousands of people all around the world strive to do.
There are numerous books and online resources on how
to plant a butterfly garden, but it all comes down to two
basic things: host plants and nectar plants.

A huge number of flowering plants provide nectar
for butterflies, but relatively few are butterfly magnets.
One that is familiar is the flowering bush buddleia
(*Buddleja davidii*), sometimes known as the butterfly bush.
In some regions it is considered a noxious plant because
of its weedy nature, but it is an undoubted butterfly
attractant. A responsible gardener, however, should avoid
planting butterfly bush outside of its native range because
it can easily become invasive. Other highly attractive
flowers for butterflies include asters, blanket flowers,
ceanothus, cone flowers, goldenrods, ice plants, knapweeds,
lavenders, lilacs, marjoram, milkweeds, phloxes, sunflowers,
teasels, thistles, valerians, verbenas, and wallflowers. Lantana
(*Lantana camara*) is extremely attractive to butterflies and
easy to grow, but should be avoided, as it has become an
invasive weed in many parts of the world.

Flowers can provide important refueling stops for
butterflies flying through urban areas, but if host plants
are also present, butterfly breeding and residence can
occur. The availability of butterfly host plants in gardens,
parks, rights-of-way, golf courses, and cemeteries has the
potential to enable butterflies to persist and thrive in the
urban landscape.

URBAN SPECIALISTS IN NORTH AMERICA

The largest butterfly in western North America, the magnificent Two-tailed Tiger Swallowtail (*Papilio multicaudata*; see pages 54–55), is a common resident of urban areas west of the Rocky Mountains. This is because one of its host plants, green ash (*Fraxinus pennsylvanica*), is a common shade tree in many towns. The equally magnificent and large Western Tiger Swallowtail (*Papilio rutulus*) is a common butterfly in the city of Seattle, again because its host plants, maples and plane trees, are commonly grown on city and suburban streets.

Another North American butterfly that can exploit urban habitats is the wide-ranging Monarch (*Danaus plexippus*). Milkweeds (*Asclepias* spp.) have become popular garden plants in recent decades, and in addition to attracting a wide range of butterfly species with their nectar, they are the sole host plants for Monarch caterpillars. Urban habitats have become an increasingly important resource for Monarchs in the USA as native milkweed habitats have declined.

Butterflies of farmland

Industrial-scale agricultural expansion over recent decades has led to the destruction and fragmentation of butterfly habitats. However, new awareness of the benefits of native habitat restoration for the biological control of crop pests is leading to increased opportunities for butterflies on farmland.

SHARING FARMLAND WITH BUTTERFLIES

Historically, farmland has not been a good home for butterflies. The expansion of farmland all over the world has resulted in the destruction or fragmentation of natural habitats, with the associated extirpation of native flora and fauna, including butterflies. Farming also uses pesticides, rendering the environment even more unsuitable.

Since the turn of the twenty-first century, however, there has been growing interest in agricultural pest management programs that incorporate conservation biological control (CBC), resulting in less pesticide use. Adoption of CBC is usually associated with native habitat restoration to provide resources for predators and parasitoids of crop pests. Native habitat restoration also provides opportunities for butterflies to re-enter the farmscape.

Flower strips to enhance predator and parasitoid populations are increasingly being planted in or near crops in Europe and North America as part of CBC programs, and likely also serve as nectar and host plant resources for butterflies. As pesticide inputs decrease and habitat restoration strategies proliferate, farmscapes will become increasingly important and valuable habitats for butterflies in the coming decades.

BUTTERFLY-FRIENDLY VINEYARDS

Under the Beauty with Benefits program in Washington state, vineyards restore native habitats and reduce their use of pesticides to minimal levels. Results have shown that these sites host at least 29 butterfly species, compared to fewer than ten species in vineyards that have not undergone habitat restoration. Many of the butterflies in these habitat-restored vineyards are resident, supported by host plants. Some vineyards in New Zealand and Europe with low pesticide input and CBC are also being developed and proving to be good habitats for butterflies.

↙ Cultivating native plants in low-input and organic vineyards improves natural control of pests and provides resources and habitat for butterflies and other pollinators.

↓ Non-pestiferous white butterflies like this Becker's White (*Pontia beckerii*) in western North America are among the first butterflies to recolonize farmland when their mustard host plants are restored and pesticide use is minimized.

The importance of climate

To moan about the weather is human. But consider how difficult it must be for a short-lived butterfly when it can't find a mate, court, or lays eggs because the weather is too cold to fly. When a lifetime lasts just days or weeks, the weather has a huge bearing on whether an individual will be successful or not in carrying on its species.

TO BE WARM IS TO BE ALIVE

Sunshine and warm summer temperatures of 68–86°F (20–30°C) are perfect for most butterflies. Under these conditions, the key components of a butterfly's life—including gathering nectar, finding a mate, and laying eggs—proceed optimally. However, just like us, there can come a point when temperatures are too hot for butterflies.

Having the right climate conditions is vital for adult butterflies—even more so than for humans, since butterflies are "cold-blooded" and depend on warm temperatures for their life activities. Overcast skies and temperatures below 54–59°F (12–15°C) will prevent most butterflies from flying, although there are some exceptions. Some dark-colored mountain species may fly at around 50°F (10°C), occasionally even in drizzle. The very dark brown European Ringlet (*Aphantopus hyperantus*) commonly flies in cool drizzle.

As we saw with mountain butterflies (see pages 124–127), increased melanism and hirsuteness are adaptations to absorb and retain every degree of warmth available. We can also see this in temperate lowland species.

THE IMPORTANCE OF RAPID GROWTH

Temperature is equally important for butterfly eggs, caterpillars, and pupae, to allow them to develop as rapidly as possible. The less time an individual spends at a vulnerable immature stage, the less likely it will be consumed by a predator or be parasitized. Caterpillars, being mobile like adult butterflies, are able to bask in sunshine to warm themselves, hastening metabolism and growth. In high summer, the heat from the sun may be too much for some, in which case they will seek cooler shaded locations. Others, including the caterpillars of Leona's Little Blue (*Philotiella leona*), can withstand very high temperatures—in this species, the heat fast-tracks development from egg to pupa in a little over a week.

THE EFFECTS OF POOR WEATHER

Precipitation can have a significant impact on butterfly lives. Rainfall is obviously important to optimize growth of host plants, and drought-like conditions—especially in marginal environments like dry grasslands—can reduce or prevent caterpillar survival and development. However, too much rainfall in some habitats may wash caterpillars right off their host plants.

Cool, cloudy, wet weather is not conducive to butterfly activity. When they coincide with the short flight periods of many species, extended periods of such conditions can significantly impede a butterfly in finding a mate and laying eggs. Cool, wet summers are a common occurrence in the UK and parts of mainland Europe, and cause substantial temporary reductions in some butterfly populations.

↑ The caterpillar of Leona's Little Blue, which is exposed to very high temperatures, completes development in little over a week.

← Most butterflies like this Buckeye (*Junonia grisea*) have profuse hair covering the body and wing bases. This helps trap warm air, increasing body temperature and so facilitating metabolism and flight.

→ Thermal images of butterfly specimens taken in the mid-infrared spectral range demonstrate how different body parts retain heat. Variation is related to the thickness of the specimens.

COPING WITH HEATWAVES

Generally, once temperatures rise above 90°F (32°C), stress sets in, and many butterflies seek moisture and shade. As long as the period of heat is not prolonged and the habitat offers opportunities to escape from the heat, most butterflies change their behavior and survive. However, global climate change is leading to more intense and prolonged heatwaves, and their impacts on butterfly populations are becoming more pronounced.

Monarchs (*Danaus plexippus*) in western North America are frequently exposed to summer temperatures of 100–104°F (38–40°C). They respond by curtailing their activity and seeking shade, where they stay until it cools down. Normally, these bursts of heat last just a few days, but in recent years heatwaves with maximum temperatures of 107–118°F (42–48°C) have lasted for up to two weeks. In these conditions, Monarchs have been observed leaving their habitat and seeking cooler locations along rivers.

↓ Butterflies thrive when temperatures are optimal, usually 54–90°F (12–32°C), and host plant/nectar resources are available.

WIND-SHY OR WIND-BOLD?

Most butterflies do not like windy weather. Even if it is warm and sunny, the majority of wind-shy butterflies (mostly small- to medium-sized species) will stay grounded when winds are blowing. Some larger species, such as large brushfoots and swallowtails, are not deterred by high winds and may utilize wind currents in dispersal and migration. Golden Birdwings (*Troides aeacus*) (right) can soar above the treetops even in strong winds.

ANATOMY OF BUTTERFLY WINGS

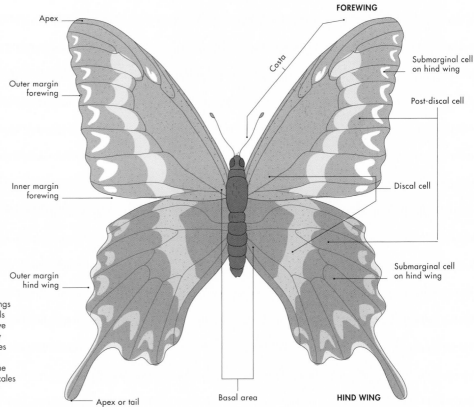

Apex

FOREWING

Outer margin forewing

Costa

Submarginal cell on hind wing

Post-discal cell

Inner margin forewing

Discal cell

Submarginal cell on hind wing

Outer margin hind wing

Butterfly wings

Butterfly wings are made from chitin (similar to fingernails) and are strengthened by air-filled veins. The wings are living structures with "blood" vessels and nerves. Some hairstreaks even have a beating heart in each forewing! They are covered with tiny overlapping scales like shingles on a roof. These scales give the wings color and patterning. The progressive and natural loss of these scales during a butterfly's life does not hinder flight ability.

Apex or tail

Basal area

HIND WING

Host plants

Along with climate, the availability of a butterfly's host plants is the most crucial factor to its survival. Without host plants and optimal weather, species will struggle to maintain their populations. They are the food of caterpillars and determine the nature of the butterfly to come.

FOOD CHOICE

Butterflies are inextricably linked to plants, not only as nectar sources for adults, but also for caterpillar food. Most adult butterflies will use whatever nectar is available, provided the timing is right and they are physically able to access it.

In contrast to the adults, caterpillars of many species have far more restricted tastes. Some require a specific family of plants, while others may feed on only a single plant species, such as Leona's Little Blue (*Philotiella leona*) on spurry buckwheat (*Eriogonum spergulinum*). Some, such as the Painted Lady (*Vanessa cardui*) and Moonlight Jewel (*Hypochrysops delicia*) caterpillars, have diverse tastes and can feed on species in many plant families.

A butterfly species that overwinters as an egg needs fresh, tender leaves as soon as it hatches in spring, so only early spring plants are candidates as hosts. Butterflies that overwinter as adults or pupae eclose as adults in the spring. These need nectar immediately, and do not require a caterpillar host plant until a few weeks later, when the plant choice will be different and more varied.

← Caterpillars of some gossamer wing (Lycaenidae) butterflies prefer to feed on host plant buds, such as this late instar Silvery Blue (*Glaucopsyche lygdamus*) caterpillar on a lupine.

↑ The caterpillars of California Sister (*Adelpha californica*) butterflies feed only on the leaves of North American oak trees.

Why do some butterflies depend on a single host plant, while others use many? There is no doubt that host plant generalists such as the Painted Lady are highly successful species, able to live in multiple habitats. However, specialization on a single host plant or a small number of closely related host plants can also be a successful strategy, as long as those plants are common and widespread on the landscape. Examples of successful specialists include the Common Evening Brown (*Melanitis leda*) and the Small Cabbage White (*Pieris rapae*), restricted to grasses and crucifers, respectively, but still both common and widespread. The strategy can become risky, however, if a butterfly becomes restricted to a host plant that is restricted itself—or suddenly becomes so.

A VARIED DIET

Some species use different host plants at different times during caterpillar development. This happens in some checkerspots (*Euphydryas* spp.), which overwinter as caterpillars. They feed on one type of plant before winter and a different plant after.

The part of a host plant preferred by caterpillars is important, and each species tends to specialize on one or two parts. The majority of caterpillars feed on leaves, but some specialize on young, new-growth leaves, while others are happy eating mature leaves. Many species, including the caterpillars of some blues, hairstreaks, and coppers (Lycaenidae), prefer to feed on leaf buds. Some of these caterpillars bury themselves into buds, burrowing into them. Other species may prefer feeding on flowers or even the seeds of a host plant.

Buds, flowers, and seeds are generally more nutritious than leaves, containing up to ten times more nitrogen than foliage of the same plant. Eating these plant parts therefore allows faster caterpillar growth, but buds and flowers may have a limited period of availability. As a result, most specialized bud and flower feeders can switch to feeding on leaves if necessary.

Caterpillars of skipper (Hesperiidae) and satyr butterflies (Satyrinae) tend to use grasses and palms as host plants. These are widely available, but they are low in nutrition. This usually means that caterpillar growth and development are slow in species that feed on these host plants.

← The caterpillars of the Echo Azure (*Celastrina echo*) preferentially feed on the flowers of their dogwood and mountain balm host plants.

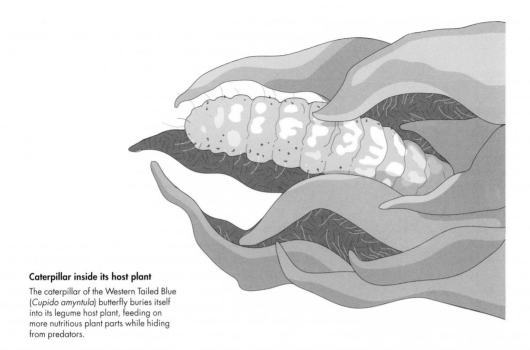

Caterpillar inside its host plant

The caterpillar of the Western Tailed Blue (*Cupido amyntula*) butterfly buries itself into its legume host plant, feeding on more nutritious plant parts while hiding from predators.

CATERPILLAR HOST GUILDS

Certain plant species and families are popular host plant choices with butterflies. For example, there is a guild of four to six butterflies in both North America and Europe that use stinging nettles (*Urtica dioica*) as a host plant, making this often reviled weed a valuable resource for butterfly rearing and conservation.

There are other examples of host plant guilds in butterflies. The genus of plants in northwestern North America commonly called buckwheats (*Eriogonum* spp.) supports a guild of 12 species of blues and coppers, plus a metalmark butterfly. Another important guild of white and sulphur butterflies (Pieridae), including the ubiquitous Small Cabbage White butterfly, favor host plants in the Brassicaceae family.

While most caterpillars feed on flowering plants, the exceptions to this pattern are fascinating. Some consume non-flowering plants, like the pine-feeding pine whites (*Neophasia* spp.), the liverwort-feeding jewelmarks (*Sarota* spp.), and clubmoss-feeding Asian ringlets (*Ragadia* spp.). The "gems" (subfamily Poritiinae) are distributed throughout the Old World tropics and, in Africa, the caterpillars of around 500 species (tribe Liptenini) consume lichens and algae, including Marshall's Acraea Mimic (*Mimacraea marshalli*; see pages 250-251). Some caterpillars feed on dead plants. For example, a few metalmarks feed on leaf litter and detritus; there is even a genus called *Detritivora*, which means "feeding on garbage." Others have given up on plants altogether. None of the harvesters and woolly legs (Miletinae) eat plants. Some feed on ant larvae—like the Moth Butterfly (*Liphyra brassolis*; see pages 94–95)—while others consume aphids and other sap-sucking insects, such as the only representative in North America, the Harvester (*Feniseca tarquinius*), which specializes on woolly alder aphids (*Prociphilus tessellatus*). The South African Peninsula Skolly (*Thestor yildizae*) spends its entire caterpillar life in the nest of pugnacious ants (*Anoplolepis custodiens*), where it tricks them into regurgitating food into its mouth until it is big enough to pupate.

Pollinating butterflies

The word "pollinator evokes" thoughts of honeybees, since these managed insects are famous for pollinating many agricultural crops. But there are also many other types of plant pollinators, including butterflies. And of all the insect pollinators, butterflies are perhaps the most beautiful.

PARTNERS IN POLLINATION

Are butterflies pollinators? Yes, they are. As frequent visitors to flowers, they invariably transfer pollen as they move from bloom to bloom to sip nectar. They are sometimes considered the number two pollinator behind bees, although their effectiveness in this role is unknown.

Butterflies are certainly less effective pollinators than most bees because their lives are less focused on visiting flowers. They also pick up less pollen than bees since their long, thin legs minimize body contact with flower parts. However, on flat, open flowers, butterfly bodies do come into contact with pollen. Pollen sticks to a butterfly's feet, legs, and proboscis, but, unlike bees, they do not have specialized structures to collect pollen.

← Large butterflies with long legs such as this Coronis Fritillary (*Argynnis coronis*) are less likely to transport pollen than smaller, short-legged species such as skippers (Hesperiidae) and gossamer wings (Lycaenidae).

↗ More research is needed to define the importance of different butterfly families in pollinating wildflowers and crops.

Pollination is a two-way partnership. The plant needs something to move its pollen to another flower, and it is willing to pay for this service. Insect pollinators don't work for free; they're paid with nectar. However, pollination will fail if the pollen doesn't land on another flower of the same kind. To increase the chance of this happening, plant species take turns blooming or secreting nectar so the number of floral choices during a single butterfly foraging trip is reduced. The architecture of the flower can also exclude certain pollinators. Few other insects have a tongue as long as butterflies and moths, and flowers with nectar at the end of long tubes are specialized for pollination by them. Moth-pollinated flowers open at night and have a strong scent, while butterfly specialists flower in the day and are frequently red— a color that honeybees cannot see.

PARARGE AEGERIA

Speckled Wood

A butterfly success story

SCIENTIFIC NAME	*Pararge aegeria* (Linnaeus, 1758)
FAMILY	Nymphalidae
NOTABLE FEATURES	Common white and brown butterfly
WINGSPAN	4½–5 in (114–127 mm)
HABITAT	Deciduous forests, gardens, and parks

The Speckled Wood is a common and successful species in European forests, where it can be found flying along sun-dappled clearings and along edges adjacent to meadows. Not deterred by dull, showery weather, it is one of the few species to fly in such conditions.

The Speckled Wood is one of the only butterfly species in the UK that is increasing in both abundance and range. This attractive spotted brown-and-white butterfly has grown in abundance by 84 percent and in range by 71 percent since the 1970s, and can now be found living in habitats beyond forests, including parks, gardens, and hedgerows. The species' extraordinary success may be a result of a warming climate, which has allowed it to reproduce more effectively and to expand its distribution to many parts of the eastern and northern UK.

The Speckled Wood is almost unique among British butterflies in that it remains active on dull, overcast days, and it also flies in shadier habitats than most other species. Adults visit flowers only occasionally for nectar, instead preferring to sip honeydew from leaf surfaces or the juice from fermenting blackberries.

A COMPLEX LIFE HISTORY

Most butterflies have just one overwintering stage, but the Speckled Wood is unusual because it can overwinter as either a caterpillar or pupa. This means there is a mixed and prolonged emergence in spring and throughout the warm months with much overlapping of generations. Consequently, Speckled Wood butterflies can be seen continuously from March to November. There are three generations annually and individual caterpillars have varied rates of development. They can sometimes have five instars (instead of four), which adds to the complexity of their life history. The green caterpillars feed on various grasses and are superbly camouflaged, as is the green pupa.

→   Populations of Speckled Wood butterflies are booming across the UK countryside, occupying non-traditional habitats and taking advantage of warmer summers.

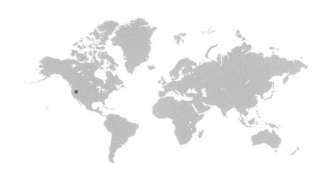

Leona's Little Blue

Tiny and rare

SCIENTIFIC NAME	*Philotiella leona* (Hammond and McCorkle, 2000)
FAMILY	Lycaenidae
NOTABLE FEATURES	One of the smallest butterflies in the world
WINGSPAN	⅝ in (15–18 mm)
HABITAT	Volcanic ash and pumice fields in a high desert

Leona's Little Blue, discovered in 1995 by Harold and Leona Rice, is confined to 15 square miles (39 sq km) of the Antelope Desert in south-central Oregon, USA, and is one of the smallest and rarest butterflies in the world.

Leona's Little Blue is a highly specialized species occupying a volcanic ash and pumice habitat in a high desert, and is dependent on a similarly specialized host plant, spurry buckwheat (*Eriogonum spergulinum*). The butterfly and its host plant occur primarily in openings of lodgepole pine (*Pinus contorta*) forest. The climate in this habitat is characterized by a short (three-month), hot, dry summer and a long, cold winter. Leona's Little Blue flies from mid-June to late July, spending much of the time low to the ground. Daily flight rarely occurs before 11 a.m. and only when temperatures are above 68°F (20°C). Most mating and egg-laying takes place during the afternoon and early evening.

A CATERPILLAR RACE

Caterpillars of Leona's Little Blue take only 10–12 days to reach the pupal stage, and usually complete development before their parents have died. This haste is necessary because the host plants senesce rapidly in the extreme summer heat of the desert habitat. The eggs, caterpillars, pupae, and adults of Leona's Little Blue are themselves exposed to direct sun temperatures of 122–158°F (50–70°C).

The tiny eggs (less than 0.5 mm in width) are laid singly at the base of flower buds. Caterpillars feed only on the unopened buds and flowers of the host plant, and are pink, red, and white in color, which provides great camouflage on spurry buckwheat's red-pink flowers. The small brown pupae are formed on the soil surface with no protection. Over winter they are exposed to temperatures as low as −22°F (−30°C), although snowfall ameliorates the extreme cold, keeping the pupae at a constant 32°F (0°C) for most of the time.

Exiting the egg
The tiny caterpillar of Leona's Little Blue eats its way out of the egg then moves to the nearest flower or bud to continue feeding.

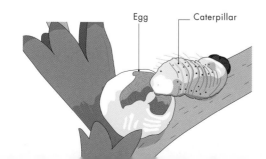

Egg — Caterpillar

→ Leona's Little Blue, discovered in 1995, is one of the world's rarest and smallest butterflies, and are confined to a tiny area of high desert in central Oregon, USA.

Ulysses Swallowtail

An emblem of the Australian tropics

SCIENTIFIC NAME	*Papilio ulysses* (Linnaeus, 1758)
FAMILY	Papilionidae
NOTABLE FEATURES	Iridescent electric-blue flashing wings
WINGSPAN	5½–6 in (140–152 mm)
HABITAT	Coastal tropical rainforests

Ulysses Swallowtails are magnificent large butterflies, immediately distinguishable by their iridescent electric-blue wings, which flash in sunlight. Emblematic of tropical Australia, the species is an iconic symbol of the antipodean rainforests.

The Ulysses Swallowtail (also known as the Blue Mountain Butterfly) was for many years a tourism emblem for Queensland, Australia, serving to entice visitors to the northern coastal tropical forests of that state.

Ulysses Swallowtail females lay their eggs singly on the young growth of their host plants, species in the genus *Melicope*. One of these, the pink-flowered doughwood (*Melicope elleriana*), has become a popular garden plant, allowing the butterfly to extend its habitat range to tropical suburban gardens. Other cultivated garden plants, including the Queensland silver ash (*Flindersia bourjotiana*) and citrus trees, are also sometimes consumed by the larvae.

The green-and-white caterpillars feed singly, preferring fresh growth. Breeding continues throughout the year, with a peak during the wet season.

Ulysses Swallowtails are an impressive sight as they fly in sunshine. The brilliant blue upperside of the wings is visible at great distances as a series of successive iridescent flashes. The butterflies may also be seen in parts of Papua New Guinea, Indonesia, and the Solomon Islands.

INSPIRATIONAL WING SCALES

Increasingly, butterflies are becoming models for advanced technological research on improving flight as well as light and heat absorption. The nanostructure of the black scales of the Ulysses Swallowtail—the most effective solar-receiving cells nature has designed—has recently inspired the creation of a high-efficiency light-absorption structure. Ultimately, this novel light-cell structure may be used in solar panels to improve their efficiency and reduce costs.

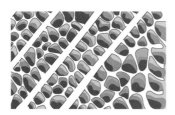

Nanostructure of a black scale

The nanostructure of a single scale from the black area of a Ulysses Swallowtail wing as seen under a scanning electron microscope. This wing structure decreases reflectivity and enhances blackness.

→ The Ulysses Swallowtail with its iridescent electric-blue flashing wings is a tropical rainforest icon and can be found in the Spice Islands, New Guinea, and Australia's Cape York.

KALLIMA INACHUS

Dead Leaf Butterfly

A butterfly that masquerades as a dead leaf

SCIENTIFIC NAME	*Kallima inachus* (Boisduval, 1846)
FAMILY	Nymphalidae
NOTABLE FEATURES	The species resembles a decaying tree leaf
WINGSPAN	3⅓–4⅓ in (85–110 mm)
HABITAT	Found in dense forests and bamboo thickets, but rarely seen in the open

Alfred Russel Wallace proclaimed this species "the most wonderful and undoubted case of protective resemblance in a butterfly," and his words still ring true today. Every aspect of its closed wings helps the insect masquerade as a leaf.

Although the Dead Leaf butterfly is closely related to the Peacock (*Aglais io*) and Small Tortoiseshell (*Aglais urticae*), it is significantly larger, and the shape of its wings is radically different. While at rest with its wings closed, the Dead Leaf Butterfly's wings have the outline of an elongated tree leaf, complete with a leaf stalk. The wing markings resemble the central vein of a leaf and the secondary veins emanating from it. Moreover, the species is highly variable. Even within a single population, different individuals resemble leaves at different stages of decomposition, complete with discoloration and irregular splotches of pigmentation resembling mold. This wing pattern variability has a complex underlying genetic basis. The different forms are maintained by evolution because predators might learn to recognize them as butterflies if they were all identical. If any form becomes so common that predators learn to recognize it,

they will begin to attack that form. This gives the other forms an advantage against predators—until one becomes so abundant that predators become wise to the trick, then this new form is preferentially attacked, etc. The species is commonly observed along forest streams and frequently feeds on rotting fruit.

Unlike the wing undersides, the wing uppersides are brightly colored with iridescent blue patches and bold, orange stripes. There are around nine species of Dead Leaf Butterfly in total—eight in addition to *Kallima inachus*. The Himalayas are a center of diversity for this group; up to four species can be found in this region. Other species are endemic to Java, Sumatra, and the Andaman Islands. Differences between species are slight, including the length and shape of the "leaf drip tips" on the fore- and hind wings plus differences in the wing upperside coloration, which are often minor.

→ Thanks to the shape of its wings, the markings that resemble a midrib, and irregular patches of "mold," this Dead Leaf Butterfly is a dead ringer for a dry, decaying tree leaf.

COLIAS NASTES

Labrador Sulphur

Living in the freezer

SCIENTIFIC NAME	*Colias nastes* (Boisduval, 1832)
FAMILY	Pieridae
NOTABLE FEATURES	Cold-zone yellow butterfly
WINGSPAN	1½–1¾ in (38–44 mm)
HABITAT	Rocky scree slopes and summits in the tundra

A specialist of alpine tundra habitats, this butterfly lives on windswept rocky scree slopes and summits. Adults fly only when the sun is out and temperatures are above 45°F (7°C). Their flight is distinctive, rapid, and low to the ground as they seek flowers and host plants in the rocky landscape.

The Labrador Sulphur is highly adapted to living in an extremely cold environment and takes full advantage of any warmth that can be derived from the landscape. Over their evolutionary history, populations of this yellow butterfly have evolved a varying density of black wing scales, which overlap and interfuse with its yellow scales. Darker individuals are more active and move larger distances because of their greater ability to achieve and maintain higher body temperatures in the brief sunshine of the Arctic summer.

Labrador Sulphurs are circumpolar in distribution and are found in tundra habitats in North America, Europe, and Russia, but they are never abundant. Both sexes visit flowers, including asters and fleabanes, and females may be seen hovering over their legume host plants, which include various tundra milkvetches.

LIFE AS A TUNDRA CATERPILLAR

Female Labrador Sulphurs lay their eggs singly on the flowers and leaves of their milkvetch host plants. The eggs hatch within a few weeks and the caterpillars begin feeding, but they soon stop as the short Arctic summer ends and the host plant senesces. Soon, the caterpillars are hibernating under deep snow. This may last for seven to nine months, before the Arctic summer returns and feeding recommences. However, in most cases the second summer is not long enough for the caterpillars to complete their development and they are forced to overwinter for another seven to nine months as almost mature caterpillars. A few weeks after waking up from their second winter slumber, the caterpillars pupate on the host plant. The adult butterflies emerge in mid-summer, to begin the cycle again.

→ The Labrador Sulphur is adapted to living on the tops of mountains and in Arctic conditions with a year-round climate that varies from cold to extremely cold.

BUTTERFLY POPULATIONS

From individuals to populations

Butterflies live their life as part of a reproducing population in a specific geographic region. The population may be small or large, diffuse or aggregated, and either expansive, occupying entire continents, or restricted to a single location. Butterfly populations are defined by the pressures they face, and how they cope with these challenges.

PREDICTABLE POPULATIONS

Individuals of a single butterfly species that live within a specific habitat or geographical area make up a population. All butterflies that could potentially mate and interact on a day-to-day basis are considered to be members of the same population. To sustain itself, a population must overcome challenges, reproduce, and adapt to minor changes in the environment. Many factors affect the health of a butterfly population: the availability of host plants at the correct developmental stage; the presence of nectar plants; the abundance of predators and parasitoids; the presence of fungal, viral, and bacterial diseases; and extreme weather events. Variability in any of these factors can have domino effects on the species' abundance.

↗ A group of male Rajah Brooke's Birdwings (*Trogonoptera brookiana*) from a Malaysian population gather at a puddle to forage for sodium.

← The Mourning Cloak (*Nymphalis antiopa*) is one of the longest-lived butterflies, sometimes living as an adult for up to one year.

For example, if a bumper crop of host plants leads to high abundance, the increased density makes it easier to transmit disease, possibly leading to an outbreak that could cause populations to plummet.

Species in tropical rainforests may reproduce continually throughout the year, but areas with pronounced dry or cold seasons may have their life stages synchronized. In temperate zones, for example, most species produce only one or two generations per year. This means that all the hungry caterpillar mouths need to be fed at the same time. If a species' eggs hatch a few weeks too late or too soon in the early summer, there might not be enough foliage to go around. Global climate change is pushing plants to start growing earlier each year. If the hatching date of overwintering eggs can't keep up, big problems are on the horizon.

BOOM AND BUST

Butterfly populations usually cycle, meaning that they may routinely increase, decline, and then increase again. The reasons for this may not be obvious and will differ according to species, the habitat, and the environment.

While predation and other mortality factors may be important, it is often the so-called abiotic factors that have the biggest impact on butterfly populations. These include the weather and availability of nectar or host plants, as well as physical space and the structure of the habitat.

Populations of Painted Ladies (*Vanessa cardui*) in North America and Europe show a regular boom-and-bust type of population cycle. In some years they are extremely abundant in northern regions such as Canada and the UK, while in others they are scarce. This appears to be a consequence of environmental conditions in their winter breeding sites in North Africa and the desert southwest of the USA. In wet winters on their breeding grounds, with plenty of host and nectar plants, Painted Ladies produce large populations that then migrate northward. Conversely, dry winters result in few northward migrants.

← The Firerim Tortoiseshell (*Aglais milberti*) roams far and wide across the landscape in North America seeking patches of stinging nettles on which to lay eggs.

→ When Lemon Emigrants (*Catopsilia pomona*) become common, they can disperse en masse.

UNDERSTANDING BUTTERFLY POPULATIONS

Many butterfly populations are transitory. This may be because resources such as nectar and host plants wax and wane. These species tend to be wider countryside species, ranging long distances, such as the Small Tortoiseshell (*Aglais urticae*) and Peacock (*Aglais io*) in Europe, and the Firerim Tortoiseshell (*Aglais milberti*) and West Coast Lady (*Vanessa annabella*) in western North America. Butterflies with long-lived perennial host plants, such as most fritillaries, blues, and hairstreaks, are more likely to have stable long-term populations than species that rely on weedy annual species.

Because a butterfly population can have so much natural variation in size from year to year, it is difficult to determine whether it is declining, increasing, or stable without many years of observation and data collection. This has become very relevant as we worry about whether species are declining or not.

Within the past few years, a growing number of scientists have concluded that we are undergoing an insect apocalypse. The abundance of all insects—not just butterflies—seems to have been declining for decades. The problem is that most researchers in the past were concerned primarily with insect diversity, not abundance, so there are few old datasets on insect abundance that can be compared with new data to track long-term trends. Population studies conducted over multiyear time frames are needed to provide insights into a range of population parameters.

An abundance of butterflies

The abundance of butterflies is determined by many factors. Suitable climate and resources, primarily nectar and host plants, are essential. But when is the last time you saw a plant completely stripped of its leaves by voracious caterpillars? With rare exceptions, natural enemies such as predators, parasitoids, and pathogens prevent numbers of caterpillars and adults from exploding. These factors probably keep butterflies from competing with each other, because resources, such as host plants, are rarely limiting.

LIMITS TO ABUNDANCE

Commonness and rarity are difficult to assess. Some species may be truly uncommon, with individuals widely separated in space or time. Others may only appear to be uncommon because they are not being counted accurately. For example, many Neotropical metalmarks rarely leave the canopy to venture down to the forest floor. Thus, few individuals wander into an observer's gaze or a collector's net. The species might be quite numerous, but the inaccessibility of its microhabitat makes its true numbers anyone's guess.

Some species are considered to be "locally common." For example, caterpillars of the Common Imperial Blue (*Jalmenus evagoras*; see pages 246–247) in Australia feed on a limited number of wattle species (*Acacia* spp.) and must be protected by the right kind of meat ants (*Iridomyrmex* spp.). Where these two resources co-occur, wattle trees can be covered with dozens of larvae and pupae that are seething with ants. The adults can form miniature clouds. But walk five meters (or 100 kilometers) in any direction and you likely won't see another until you happen upon another spot with the right combination of resources.

↗ The California Tortoiseshell (*Nymphalis californica*) is a boom-and-bust butterfly that sometimes aggregates in the thousands on damp ground in the mountains of western North America.

← Undisturbed natural habitats sometimes allow large populations of butterflies to develop. When butterflies gather to absorb salts and minerals from damp ground, such as these checkerspots (*Euphydryas* spp.) and blues (Lycaenidae), spectacular aggregations occur.

IRRUPTIONS

The Painted Lady (*Vanessa cardui*) is the most famous example of an irrupting butterfly in North America and Europe. When an irruption occurs in North America, the vast numbers of migrating Painted Ladies can become a traffic hazard, with the slippery mess of tens of thousands of squashed butterflies sometimes even causing road closures.

The California Tortoiseshell (*Nymphalis californica*) is another western North America butterfly prone to population booms and busts. Unlike the Painted Lady, whose populations appear to respond to good winter breeding conditions, we do not know why California Tortoiseshells (another migratory species) sometimes produce enormous populations and then virtually disappear. Its population booms do not appear to be related to climate, so it is possible they are connected to reduced pressure from predators, parasitoids, or pathogens. Related species such as the Mourning Cloak (*N. antiopa*) and Compton Tortoiseshell (*N. l-album*) also go through similar boom-and-bust cycles, and these also may be related to varying degrees of pressure from natural enemies.

Another group of butterflies prone to population booms is the Lemon Emigrant (*Catopsilia pomona*), in Asia and Australia. The caterpillars feed on a wide variety of legumes, including ornamental *Cassia* trees. When a large cohort of caterpillars matures simultaneously, clouds of yellow butterflies can be seen moving en masse in the same direction, justifying their common name.

Another blizzard-forming butterfly is the Pine White (*Neophasia menapia*; see pages 182–183), native to the Pacific Northwest of North America, which sporadically can become as abundant as snowflakes in a winter snowstorm. Millions of these weak-flying white butterflies sometimes decorate vast areas of pine forest in Oregon and Washington, and may get lifted by air currents and transported hundreds of miles away to areas with no pine forests. Outbreaks of Pine Whites may occur only every 30 or 40 years, and in most years their populations are small.

→ Aggregations of colorful butterflies on damp ground, such as these Gaudy Altinotes (*Altinote negra*), are a common sight in tropical regions.

(Inset) Blue Tiger (*Tirumala limniace*) butterflies gather and aggregate in large numbers prior to and during migrations in parts of Asia.

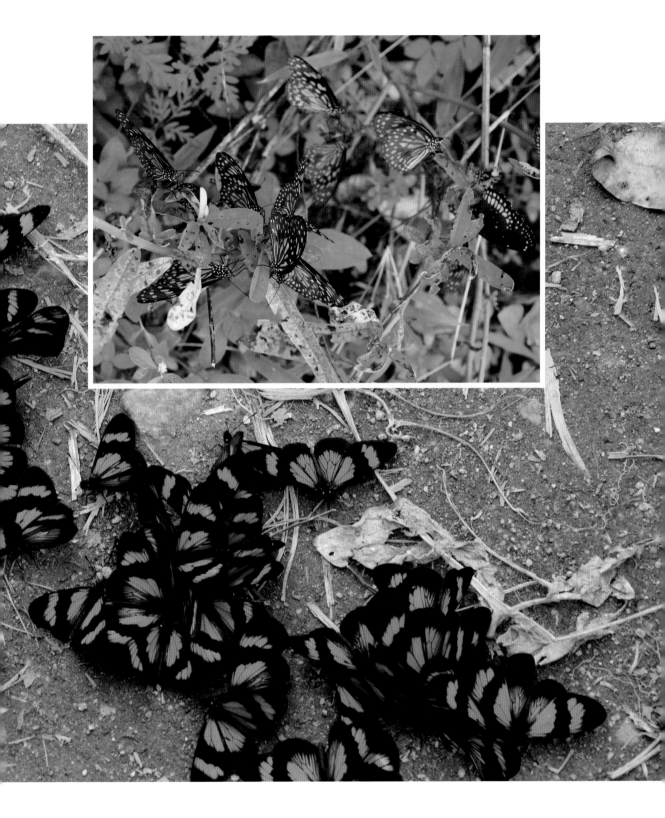

Metapopulations

Metapopulations are populations of populations. Understanding the biological basis of these networks is critical to studying the ecology, evolution, and conservation of butterflies. A butterfly metapopulation consists of discrete populations scattered across the landscape, all of which are usually linked, meaning that individuals have the potential to move between them.

← Organic or low-input croplands can be made "friendlier" to butterflies by using ground cover that consists of inter-row flowering that provides nectar and host plants.

THE IMPORTANCE OF CONNECTION

A butterfly population cannot exist in isolation forever. It may survive for a while, but ultimately it will need genetic diversity to prosper. For this to happen, individuals need to be able to move from one population to another. The sustainability of the whole metapopulation of the species across the landscape is therefore critical for effective conservation.

In a metapopulation, movement between populations is far less common and more difficult than movement within populations. This spatial organization has several benefits. For example, a virulent disease may spread quickly, wiping out all of the individuals within a population and leaving their territory uninhabited. The difficulty of moving between populations safeguards the others from infection. Later, a wandering, mated female might lay eggs and repopulate the area where the extinct population used to live, perhaps from a population with stronger immunity.

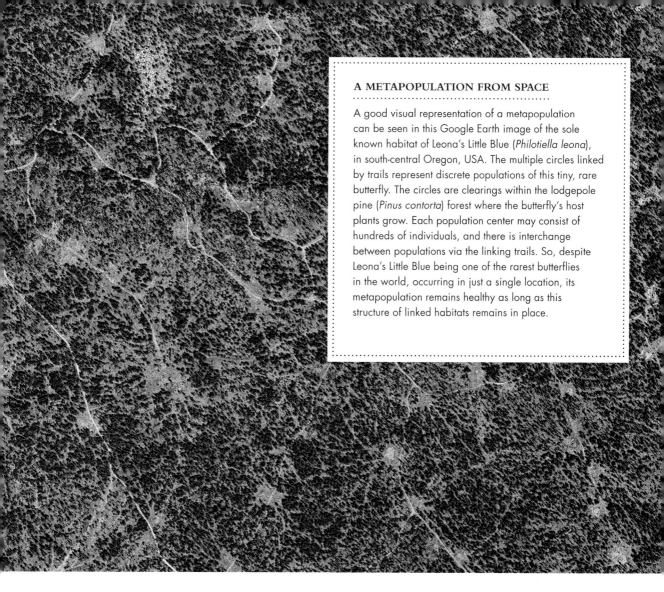

In agriculture-dominated landscapes, as now occur throughout much of Europe, butterfly populations may exist within small areas of natural habitat among a sea of cropland. As long as individual butterflies can travel between these natural habitats through the matrix of crops, the separate populations can survive. If the islands of natural habitat are large, they will support more robust populations than if they are small.

Small populations of butterflies in the fragmented landscapes of Europe and heavily cropped areas in North America may be extinguished. However, if populated habitats can be linked to other flourishing populations, then they have the potential to persist.

Large habitat patches can support large populations, while small patches support smaller populations. However, in isolation, inbreeding or rare disasters doom every population to extinction. Human agriculture and development have fragmented natural landscapes throughout most of the planet. Connecting the remaining patches with corridors of suitable habitat is crucial for maintaining healthy, viable metapopulations. Depending on the species, these corridors might include hedgerows, gullies, or tracts of forest.

Butterfly migration

Some butterflies are great fliers; others not so much. Long-distance movement is essential for the vitality of a metapopulation network. Such movement is typically called "dispersal," and helps a species spread to greener pastures. When animals make long-distance round-trips, it is termed "migration."

LONG-DISTANCE MIGRANTS

Many of the birds seen during the North American or European summer fly to the opposite hemisphere when temperatures drop. A few butterflies also undertake similar long-distance migrations, but there is a crucial difference that explains why true migration is less common in insects. Birds live for several years, and young adults learn the route from their older relatives. In butterflies, there is no possibility of learning the route because no individual makes the journey more than once. In fact, most butterflies don't even make the whole round-trip, as it usually takes several generations to complete the journey. The butterflies that fly south for the winter die, and their offspring (or their offspring's offspring) fly north the next spring. This means that the entire route and all the navigational tools needed to travel hundreds of miles must be genetically (or epigenetically) encoded. Moreover, since it might take two or more generations to complete

← Western populations of the American Monarch (*Danaus plexippus*) migrate from Canada and the USA west of the Rocky Mountains to California where they overwinter in up to 400 coastal sites.

↗ Milkweed butterflies in Taiwan, including these Blue Tigers (*Tirumala limniace*), migrate to the southern tip of the island to spend the winter.

the round-trip, a newly eclosed butterfly needs to "know" what part of the circuit it is in to choose the correct flight path (or to stay put). Some summer generations of Monarch butterflies do not migrate and live much shorter lives than migratory generations.

The really great butterfly travelers are renowned for their long-distance migrations spanning countries and continents. The most famous of these are the Monarch (*Danaus plexippus*) and the almost cosmopolitan Painted Lady.

Both of these butterflies fly around 1,800–3,100 miles (3,000–5,000 km) in the fall to reach their overwintering areas. The Monarch in eastern North America flies from Canada to mountains in central Mexico for overwintering. In Europe, the Painted Lady flies from the UK and northern parts of the continent to spend the winter months in sub-Saharan Africa. Monarch butterflies can see polarized light and sense

the Earth's magnetic field to help them orient. In addition, they use the sun as a navigational compass, automatically adjusting for its movement across the sky each day. Both species fly at high altitudes for at least part of the journey. Glider pilots have seen migrating Monarchs heading south at an elevation of 10,800 ft (3,300 m), while radar in Europe has detected Painted Ladies at 1,600–3,300 ft (500–1,000 m) taking advantage of wind currents. While the details of Painted Lady migrations are still being worked out, we know that Monarchs return to ground level each evening to feed on nectar and roost, sometimes in aggregations.

Butterfly migration is not confined to Painted Ladies and Monarchs. Migration in many other species is less well known, however, often because their movements are less dramatic. Red Admirals (*Vanessa atalanta*) in Europe and North America appear to migrate, particularly in spring. At this time of year

MONARCHS IN WESTERN NORTH AMERICA AND AUSTRALIA

Monarchs also occur west of the Rocky Mountains. Based on tagging conducted in 2012–2022, researchers have found that this population migrates around 600 miles (1,000 km) to numerous overwintering sites in coastal California. Some of these sites, such as those at Pismo Beach and Pacific Grove, are famous tourist attractions. While the California populations are substantially smaller than the overwintering populations in Mexico, some sites can host up to 20,000 butterflies on just a few trees, providing an awe-inspiring sight.

Monarchs arrived in Australia in 1870, dispersing and island-hopping across the Pacific from North America, and soon adapted to local conditions. In subtropical Australia they breed year-round, but in more temperate regions such as New South Wales they have evolved migration and overwintering behaviors that are slightly different from those seen in North American populations. Monarch migration in Australia happens on a smaller scale, covering a few hundred miles, and overwintering occurs mostly in the Sydney Basin. These behaviors are more flexible than in North America, because of the generally warmer climate in Australia and the reduced need to escape inland areas in winter.

they sometimes suddenly turn up in large numbers in northerly areas, suggesting a migratory influx. Milkweed butterflies (tribe Danaini), which includes the Monarch, are especially prone to migration. For example, the Chestnut Tiger (*Parantica sita*) routinely travels between Japan and Taiwan over hundreds of miles of open ocean. Within Taiwan, multiple crow butterfly species (*Euploea* spp.) migrate to the southern tip of the island to spend the winter in large groups, much like the Monarchs in Mexico.

THE PURPOSE OF MIGRATION

Migrating butterflies are the jet-setters of the insect world. Unlike human jet-setters, however, their travel is essential to their survival. They seek better temperatures so they can reproduce successfully, and to search for necessary food resources for their offspring. Painted Ladies and Monarchs differ in what they

do once they reach their overwintering areas. Once Painted Ladies arrive in sub-Saharan Africa, they begin to breed opportunistically, producing one or two generations from December to February. Then, once days start to lengthen, populations begin migrating north. Unlike the southerly migration, this northward migration is completed by a number of overlapping generations.

In contrast, Monarchs spend their winter in the Mexican highlands as a semi-dormant, non-reproductive population, clustered in countless millions on oyamel fir (*Abies religiosa*) trees. Like Painted Ladies, they use a multigenerational strategy to recolonize the USA and Canada in spring. Individuals that make the entire return trip of up to 4,350 miles (7,000 km) are exceptionally rare.

← By tagging Monarch (*Danaus plexippus*) butterflies, scientists have developed a good understanding of the ecology of migration of this iconic species in North America and Australia.

→ The Common Crow (*Euploea core*) and related butterflies migrate between the east and west coasts of India in response to monsoonal rains.

RADAR AND TAGGING

Our understanding of these incredible butterfly migrations has been derived in a few different ways. The dynamics of Painted Lady migrations have only recently been worked out from citizen science and radar observations, while decades of tagging Monarchs with small sticky labels has been very successful in providing data on individual butterflies and their movements.

Painted Ladies also occur in North America, and similar seasonal migrations occur on that continent, but the details of these are not as well known. Winter breeding occurs in Mexico and the southern USA, and in some years large populations are seen migrating northward.

ALTITUDINAL MIGRATION

Some brushfoot butterflies—including the Mourning Cloak (*Nymphalis antiopa*) and Firerim Tortoiseshell (*Aglais milberti*) in western North America—undergo altitudinal migrations. These migrations appear to occur in response to warming temperatures during spring, when these species will suddenly disappear from lowland areas and reappear in high-elevation mountain meadows. There is also evidence for a downslope movement of these butterflies in fall.

One of the best examples of altitudinal migration is that of the Coronis Fritillary (*Argynnis coronis*) in Washington state. This butterfly feeds as a caterpillar in spring on sagebrush violets (*Viola trinervata*) in its shrub-steppe habitat, and adults emerge in May. By June, the shrub-steppe has become sunny and hot,

←← The Caper White (*Belenois java*) undertakes long-distance movements in eastern Australia, but the reasons for this have yet to be discovered.

← The Clouded Yellow (*Colias croceus*) migrates annually from North Africa and southern Europe to the UK and northern Europe. Like many migratory species, it can be gregarious, with migrants arriving like a yellow cloud off the sea.

→ The Western Pygmy Blue (*Brephidium exilis*) is one of the world's smallest butterflies, yet is still capable of migrating from southern to northern areas of the USA each summer.

and the landscape dries out. At this point, Coronis Fritillaries take to the wing and migrate to the Cascade Mountains, about 60 miles (100 km) away. They then spend July and August sipping nectar in mountain meadows. At the end of August, they migrate back to the shrub-steppe and lay their eggs where the sagebrush violets will grow in the spring.

MYSTERIOUS EMIGRATIONS

Mass movements occur in most families, but the behavior is largely unstudied in many species. For example, in Australia, clouds of flying Caper Whites (*Belenois java*) are periodically seen. However, the details of how and why they are emigrating, where they came from, and where they are going are sketchy.

Presumably, as with most migrating butterflies, the reason for their migration has something to do with avoiding unfavorable weather or seeking new host plant resources.

Even the tiny Western Pygmy Blue (*Brephidium exilis*) in North America, one of the smallest butterflies in the world, migrates north every summer from its year-round breeding areas in the south. How individuals of this species migrate is unknown, but, given their size, it seems likely that they are assisted by high-altitude prevailing winds.

↑ Millions of Monarch butterflies (*Danaus plexippus*) overwinter at high elevation sites in central Mexico by roosting on fir trees. However, on sunny days they may take flight and fill the sky.

↗ Caterpillars of Caper White (*Belenois java*) butterflies denude their caper tree host plants and the newly eclosed butterflies take to the air to begin their mysterious migration.

Painted Lady

Pretty jet-setter

SCIENTIFIC NAME	*Vanessa cardui* (Linnaeus, 1758)
FAMILY	Nymphalidae
NOTABLE FEATURES	Salmon-pink to orange butterfly with black-and-white markings
WINGSPAN	2¾–3 in (70–75 mm)
HABITAT	Almost anywhere, from cities to mountaintops

Long-distance migration is a big part of the Painted Lady's life cycle. The species is found on all continents except Antarctica, although its range is limited in South America and Australia.

Every fall, the Painted Ladies in northern Europe travel extraordinary distances of up to 3,100 miles (5,000 km) to sub-Saharan African countries including Chad, Benin, and Ethiopia. These flights are fueled by nothing more than nectar and wind, are guided by a brain the size of a pinhead, and are facilitated by wafer-thin wings made from hardened protein. Parts of these migratory flights even take place during darkness.

One or two generations of caterpillars are produced in Africa, before new adults begin heading north. This northward movement is achieved by multiple generations of butterflies, which eventually reach northwestern Europe by early summer. Little is yet known about the Painted Lady's migration in Asia and other parts of the world.

NOT A PICKY EATER

Female Painted Ladies look for thistles to lay their eggs on, but they are not too fussy, knowing that their caterpillars will feed on at least 100 plant species. Thistles, asters, daisies, and mallows are favored host plants, but occasionally Painted Ladies will lay their eggs en masse in soybean fields and their caterpillars will then feed on the legumes, causing economic damage. This ability to use a wide range of host plants is a key reason for the worldwide success of the Painted Lady. Caterpillars will even feed on artificial diets that have been created from agar and ground-up mallow leaves.

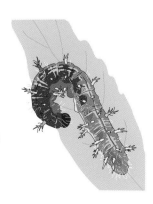

Painted Lady caterpillar
Painted Lady caterpillars feed on many different plant species including legumes, lupines, mallows, nettles, and thistles, as well as cultivated species such as soybeans.

→ The Painted Lady, one of the butterfly world's premier jet-setters, is present on all continents except the Antarctic.

DANAUS PLEXIPPUS

Monarch

The poster child of butterflies

SCIENTIFIC NAME	*Danaus plexippus* (Linnaeus, 1758)
FAMILY	Nymphalidae
NOTABLE FEATURES	Striking black, orange, and white butterfly
WINGSPAN	3–4 in (76–102 mm)
HABITAT	Almost any open habitat, especially riparian

The butterfly chosen to illustrate a children's book, to advertise a product, as a tattoo, or to symbolize marriage, birth, or death will invariably be a Monarch. The species is indeed the world's butterfly.

The Monarch has achieved its fame largely through its spectacular transcontinental migration from eastern North America to Mexico every fall. The destination of these migrants was a mystery until 1976, when millions of the butterflies were found roosting during the winter on oyamel fir (*Abies religiosa*) trees in the highlands of central Mexico and a photo of the amazing discovery made the cover of *National Geographic*. We now know that nearly all of the Monarch population from eastern North America covers all the trees in an area of 7–44 acres (3–18 hectares), and it has been estimated that there may be 50 million Monarchs for every 2.5 acres (1 hectare) of covered trees. Unfortunately, the number of overwintering Monarchs in Mexico has been declining because of habitat loss, climate change, and logging in this unique area. Smaller migratory populations of Monarchs exist in the western USA and in New South Wales, Australia.

In both cases, the butterflies migrate from inland areas to overwinter on trees, often *Eucalyptus*, near the coast. Much genomic and neurobiological research has focused on uncovering the basis of this amazing insect's internal system of orientation and navigation.

EATING POISON FOR PROTECTION

Monarch caterpillars need milkweeds (*Asclepias* spp.) to feed on as a caterpillar. Milkweeds' milky sap contains toxic cardiac glycosides—steroids that disrupt the flow of sodium and potassium ions in and out of cells, and are toxic to most animals, including other insects. Monarchs and their close relatives are largely immune to the effects of the compounds because they have a handful of mutations that prevent the cardiac glycosides from affecting them. Caterpillars feeding on milkweeds accumulate these poisons, making them toxic to potential predators. They advertise their toxicity and distastefulness by their bold black, yellow, and white stripes. The poisons are carried over to the adults, and the striking black, orange, and white coloration of the butterflies is a warning to their enemies: eat me and I may kill you.

→ The Monarch butterfly is the world's best-known butterfly, featured in many aspects of human life including books, advertisements, television, and films. It is also a very popular subject for tattoos.

Common Palmfly

A remarkable "dual mimic"

SCIENTIFIC NAME	*Elymnias hypermnestra* (Linnaeus, 1763)
FAMILY	Nymphalidae
NOTABLE FEATURES	Geographic variation in mimicry
WINGSPAN	2–2¾ in (50–70 mm)
HABITAT	Lowland forest up to 3,300 ft (1,000 m) and open areas

Common Palmflies are found throughout the Asian tropics and subtropics. Depending on local circumstances, females can mimic one set of toxic species or another with the flip of a genetic "switch."

Nearly all of the palmflies are perfectly edible Batesian mimics that resemble toxic "model" species. The aptly named Common Palmfly is the most abundant and widespread species in this group, and is considered a "dual mimic" because each sex can mimic a different model. Males mimic dark crow butterflies (*Euploea* spp.) and females often mimic orange tiger butterflies (*Danaus* spp.). However, the situation is even more complex for the Common Palmfly. In some places, the females also mimic crow butterflies, just as the males do. In India, Thailand, and Java, for example, females are orange. In Borneo, Taiwan, and the Lesser Sundas, they are dark. The switch between orange or dark females appears to be determined by just two DNA bases in the entire genome. It is likely that these DNA nucleotides act as a switch for turning one set of genes for turning on dark female wings or another set of genes for orange female wings.

AIRBORNE MIMICRY

Remarkably, only the upperside of the wings seem to be mimetic. The underside of the wings are mottled and resemble tree bark. While at rest, perched on the trunk of a tree, the wings are opaque and the butterfly blends into the background. However, when it flies, the wings are so thin that pigments on the upperside of the wing shine through to the underside. Therefore, the Common Palmfly seems to be a mimic while in flight (when seen from above or below), but camouflaged while at rest.

The caterpillars feed on a wide range of palms, including ornamental species used in tropical city landscaping. The adult butterflies are never far from palms and can be found in some of the most densely populated Asian cities. The adult butterflies prefer to feed on rotting fruit, which can bring them out of hiding in a dense palm thicket.

→   A Common Palmfly rests on a tree in Singapore. All males of this species resemble toxic crow butterflies (*Euploea* spp.), and sometimes females do, too.

AGLAIS IO

Peacock

Wings like peacock feathers

SCIENTIFIC NAME	*Aglais io* (Linnaeus, 1758)
FAMILY	Nymphalidae
NOTABLE FEATURES	Stunning large eyespots on each wing
WINGSPAN	2¼–2½ in (55–63 mm)
HABITAT	Gardens, grasslands, parks, and open forests

The Peacock butterfly is well named for the single eyespot on the upperside of each wing, which resembles the markings on a peacock's tail feather. These bright eyespots are visually jarring when the wings are briefly and suddenly opened, and are an effective ploy for deterring bird predators.

The Peacock is a bold, stunning beauty and one of the few butterfly species that benefits from drier and warmer summers. It has extended its range northward and as a consequence is now common at higher latitudes across its native European range. This success is enhanced by the fact that its host plant, the stinging nettle (*Urtica dioica*), is common and also doing well.

SCARING RODENT PREDATORS

Late-summer Peacocks spend a lot of time feeding on nectar and the fermenting sugars of fallen fruit, converting them to fats that they then store to provide energy later. This is a good plan, because for three to four months over the winter the butterflies retreat to a cool, dark place such as a cellar, garage, shed, or tree hole. As in hibernating bears, the stored fat will enable the overwintering Peacocks to survive. However, danger lurks in the dark, usually in the form of hungry rodents looking for a fat- and protein-rich snack. Although the jet-black lower wing surfaces of the Peacock make it invisible in the dark, the hungry rodents can smell their snack. However, a sleeping Peacock has some scary tricks up its sleeve, including wing-flicking, ultrasonic clicks, and hissing. This sonic and ultrasonic assault usually does the trick and the spooked rodents flee.

Hibernation
The Peacock butterfly hibernates in winter, often choosing the ceiling of a shed or outhouse. Such a place provides protection from the worst of winter weather, sheltering the butterfly from the wind and cold temperatures.

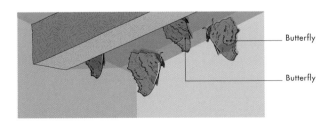

Butterfly

Butterfly

→ The Peacock butterfly is well-named with the large eyespots of its namesake on each wing, which it uses to flash and scare naïve birds and rodents.

NEOPHASIA MENAPIA

Pine White

White butterflies of pine forest

SCIENTIFIC NAME	*Neophasia menapia* (Felder & Felder, 1859)
FAMILY	Pieridae
NOTABLE FEATURES	White butterfly prone to occasional irruptions
WINGSPAN	1¾–2 in (45–50 mm)
HABITAT	Pine forests

The Pine White is a boom-and-bust butterfly. In most years it is barely noticeable, with just one or two visible in pine forests, slowly meandering through the treetops. Then, every three or four decades, the population irrupts and transforms into a blizzard.

The Pine White is a northwestern American butterfly, occasionally seen flying in ones and twos around the tops of trees in its pine forest habitat. However, every 30 or 40 years something happens. The normal mortality pressures that ensure only a fraction of Pine White eggs make it to adulthood disappear. Suddenly, nearly all the eggs that are laid develop into caterpillars, which then turn into butterflies—and a snowstorm is born. Millions upon millions of Pine Whites fill the forest and sky. Countless numbers are lifted high up into the atmosphere and hitch a ride on the winds, sometimes coming down in places where no pine trees exist. This population explosion continues for two or maybe three seasons, then, as suddenly as it arrived, it disappears. In its wake, thousands of pine trees stand without their needles, all devoured by the hungry Pine White caterpillar hordes. Luckily, most trees survive this onslaught, and then remain unbothered by Pine Whites for another few decades.

NEEDLELIKE CATERPILLARS

Few butterfly caterpillars can feed on conifers. The Pine White and its sister, the Chiricahua Pine White (*Neophasia terlooii*), are the only members of their family (Pieridae) anywhere in the world that can eat anything other than flowering plants. Moreover, Pine White caterpillars are masters of disguise, looking like the pine needles they rest on. This amazing camouflage makes them very hard to find, although perhaps some birds have learned to detect them. More likely potential predators are insects such as stink bugs, which find their prey using smell and sound. One thing Pine White caterpillars and eggs cannot hide from are pathogens such as bacteria, viruses, and fungi. Perhaps it is these that keep such a tight rein on Pine White populations under "normal" conditions?

→ The Pine White flies gently and softly around the tops of pine trees in most years, scarcely noticed, until the time when the population becomes a snowstorm.

BUTTERFLY
SEASONALITY

Hibernation: surviving the winter

Butterfly hibernation is no less impressive than the hibernation of bears. By changing their physiology, butterflies or their eggs, caterpillars, or pupae are able to survive the winter in a state of suspended animation.

HIBERNAL DIAPAUSE

The majority of the world's butterflies are tropical and never experience cold temperatures. The ability to withstand low temperatures inhospitable to insect life is a key innovation enabling some butterflies to survive in temperate regions. For example, nearly 170 species of milkweed butterfly can be found in tropical areas around the world. Although they have managed to circumnavigate the globe, none of them have evolved the ability to withstand freezing temperatures. Only one milkweed butterfly species can live in temperate climates, and it does so by traveling hundreds of miles every year to avoid winter: the Monarch butterfly (*Danaus plexippus*). However, most of the butterflies that can live in areas of the world where it freezes annually can survive from year to year by hibernating. In insects, hibernation is a physiological state technically called diapause. This is characterized by a lowered metabolic rate and radical biochemical changes. It is different from the simple dormancy or inactivity that occurs in butterflies during cool periods or overnight. Diapause is a rigidly controlled mechanism that is genetically fixed or induced by environmental conditions.

Some butterflies are genetically programmed to enter hibernal diapause at the same stage and at the same time every year. Different species overwinter in different stages, either as adult butterflies, eggs,

caterpillars, or pupae. These are usually species that occupy environments that are only suitable during a narrow time window, such as Arctic or alpine habitats, where there is insufficient time to produce more than one generation a year. For example, the diapausing eggs of Parnassian butterflies (Papilionidae) that inhabit mountain environments cannot be "persuaded" to hatch before winter, regardless of the temperatures. These eggs are programmed to remain in diapause until after exposure to cold winter temperatures for a certain period of time—usually two to eight months. Once the necessary period of chilling has occurred, a physiological "switch" is thrown that allows the eggs to develop when it is warm enough.

The same refractory mechanism occurs in diapausing adult butterflies, caterpillars, and pupae. For most species, the overwintering stage usually gains the competency to develop just after the shortest day of the year, when day lengths are increasing again.

→ The diapausing egg of a Mountain Parnassian (*Parnassius smintheus*) lies dormant under winter snow until spring days and warmth stimulate the hatching of a tiny caterpillar.

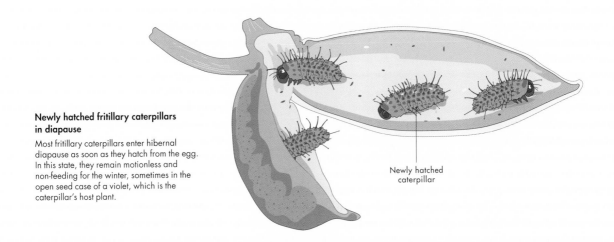

Newly hatched fritillary caterpillars in diapause

Most fritillary caterpillars enter hibernal diapause as soon as they hatch from the egg. In this state, they remain motionless and non-feeding for the winter, sometimes in the open seed case of a violet, which is the caterpillar's host plant.

Newly hatched caterpillar

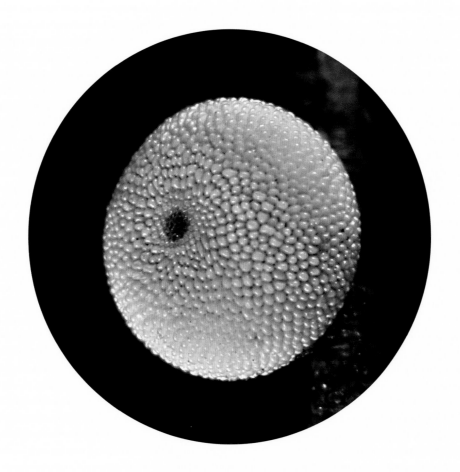

ENVIRONMENTAL CUES

Some species that have only one generation in most years do not have fixed hibernal diapause. In these species, as well as those with two or more annual generations, diapause is determined environmentally. The most important environmental cues determining the induction of hibernal diapause in butterflies are declining day lengths and cooling temperatures. Other cues, such as senescing host plants, may also have an influence in some species.

AN EXTRA GENERATION

Diapause is environmentally determined in some species. If environmental cues are suitable, butterfly species that normally have a single generation may opportunistically "sneak in" an extra generation. Day length or the rate of declining day length is the most reliable indicator of seasonal change and is the cue most often used by butterflies. If above-average spring temperatures enable a butterfly species to complete caterpillar development faster than normal, day lengths that are still increasing may signal an opportunity to complete another generation.

The Purplish Copper (*Tharsalea helloides*; see pages 204–205) in western North America is a good example of a species that may have one, two, or three generations in a single year depending on the environmental cues experienced by a population. Females that eclose in June produce non-diapause eggs and another generation of adults emerge in August. These August adults may produce diapause or non-diapause eggs, and in a long summer season the non-diapause eggs will produce

a third adult generation. If the season finishes quickly and this generation fails, the diapausing eggs from the August adults ensure the population continues into another season.

Sometimes a butterfly will, metaphorically, place all its eggs in one basket. An example is the Wall Brown (*Lasiommata megera*) in the UK. In inland areas, warmer summers in some years have tempted the normally double-brooded Wall Brown to try to produce an extra generation in the fall. In most years this fails dismally, with cold weather preventing caterpillar development. This has caused substantial population declines and extirpations in central England, where the species used to be common.

FINDING A WINTER HOME

Diapausing eggs and pupae overwinter in microhabitats chosen by female butterflies and caterpillars, respectively. Diapausing eggs are usually laid on dead host plants or on twigs, branches, or rocks.

Overwintering pupae are formed in similar places on, close to, or under the ground surface or on branches. In all cases, eggs and pupae are cryptic to minimize the chances of being detected by hungry birds or rodents.

When caterpillars enter diapause in late summer or fall, they seek out hibernacula, or protected places, where they will spend the winter. Common hibernacula include curled leaves, seed pods, and crevices in rocks or in the soil. Some checkerspot caterpillars (*Euphydryas* spp.) overwinter en masse in silken nests on the ground.

When you see an immaculate-looking Peacock (*Aglais io*) or Small Tortoiseshell (*A. urticae*) butterfly sunning itself on an early spring day in Europe, it is often hard to believe that it has been a butterfly since the previous summer. After spending late summer and the fall quietly stocking up on sugars from nectar, sap flows, and fallen fermenting fruit, this butterfly hibernates through the dark, cold winter months in a dark shed or tree hole. With minimal energy expenditure and no wear and tear, it awakens in spring looking as fresh as when it began its diapause.

Overwintering checkerspot caterpillars
Some checkerspot caterpillars overwinter on the ground in dead leaves under a silken nest or shelter they create in late summer or fall. On mild late winter days, they sun themselves on the top of the nest.

← The Wall Brown (*Lasiommata megera*) appears to have been tricked by warmer summers in the UK into producing a non-sustainable extra generation.

←← The Purplish Copper (*Tharsalea helloides*) is an opportunistic butterfly squeezing in as many generations as possible from spring to fall.

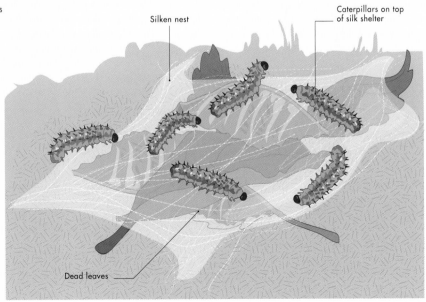

Silken nest

Caterpillars on top of silk shelter

Dead leaves

Estivation: dormancy through summer

Some butterflies find the dog days of summer just too hot, and so choose to sleep these away in a cool spot out of the sun. Some do this as adults, while others undergo summer torpor as caterpillars or pupae. And just like hot dogs, all butterflies at any stage avoid activity and development at the height of summer.

DEALING WITH THE HEAT

In places with a Mediterranean climate, surviving a hot, dry summer can be just as challenging for butterflies as surviving a cold winter. Weeks or months of daily temperatures above 95°F (35°C) and with little to no rain also reduce host plant and nectar availability, so it clearly makes sense for butterflies in these habitats to press pause for a while.

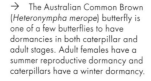

→ The Australian Common Brown (*Heteronympha merope*) butterfly is one of a few butterflies to have dormancies in both caterpillar and adult stages. Adult females have a summer reproductive dormancy and caterpillars have a winter dormancy.

While dormant butterflies in winter hibernate, in summer, they are said to estivate. Both dormancies are characterized by physiological diapause, which in summer prevents development or activity (or both) until the heat has passed. As with hibernation, changes in day length are an important cue for inducing summer dormancy, along with increasing temperatures and senescing plants. Termination may be cued by decreasing day lengths and temperatures.

In western North America, early summer eclosing butterflies that estivate include various greater fritillary species, browns (Nymphalidae), and the Mourning Cloak (*Nymphalis antiopa*; see pages 198–199). The Great Spangled Fritillary (*Argynnis cybele*) ecloses in June and the females immediately seek a resting place in the forest, "disappearing" until late August. The males remain active throughout the summer, apparently less fazed by the summer heat. The estivating females are non-reproductive and do not develop ovaries until late summer. The survival of eggs and caterpillars in the Great Spangled Fritillary is improved by waiting until the heat of summer has passed and conditions are moister.

DOUBLE DORMANCIES

The Mourning Cloak is another species that ecloses in June and then lies low for the summer, although in this case both sexes estivate. They emerge briefly in fall to feed on sap flows and fermenting fruit, before entering a second dormancy, this time true hibernation. The species is one of the longest-lived butterflies,

↑ Male Great Spangled Fritillaries do not estivate. Females estivate soon after eclosion in early summer, and remain mostly dormant until late summer when they become active and lay eggs.

↖ The long-lived Mourning Cloak estivates in the summer and hibernates in the winter.

↗ Coronis Fritillaries have a summer reproductive dormancy and migrate to higher elevations, spending two months in mountain meadows, before returning to the shrub-steppe.

likely facilitated by having these two extended dormancies. A few species, including the Common Brown (*Heteronympha merope*; see pages 200–201) of southern Australia, have an extended summer dormancy as an adult, and a winter dormancy as a caterpillar. However, diapause in two life stages of a butterfly is not common.

Some butterflies with a summer dormancy also undergo an altitudinal migration, which takes them to a cooler environment. Examples include some western North American butterflies that spend summer at a higher elevation and return to lower areas in the fall, such as the Mourning Cloak, Firerim Tortoiseshell (*Aglais milberti*), and Coronis Fritillary (*Argynnis coronis*).

Phenology: getting the timing right

Phenology is about doing the right things at the right time, especially in relation to the changing seasons. Butterflies fly when the sun shines and temperatures are warm. Phenological differences among species provide a window into how climate and unpredictability shape the lives of butterflies.

↑ The Arctic Blue (*Agriades glandon*) is an alpine butterfly with just a single generation of adults flying each year. Overwintering as a half-grown caterpillar, some individuals may take two years to reach adulthood.

↗ In coastal lowland habitats, the Anise Swallowtail may have two or three generations annually. However, in montane and arid habitats, only a single generation is possible.

LIFE HISTORIES

The meaning of phenology is neatly summed up by the biblical quote "To every thing there is a season, and a time to every purpose under the heaven" (Ecclesiastes 3:1). The life history of a butterfly—how it develops, survives, and reproduces over time in response to temperature, day length, and season—is an essay in phenology. This is a rich area for natural history study, as the details of the immature lives of even common butterflies are sometimes poorly known. There are many species whose eggs, caterpillars, and pupae have yet to be studied or photographed.

The lives of eggs, caterpillars, and pupae and how they determine the phenology of butterflies are still being unraveled. We know quite a lot about well-studied butterfly faunas, such as those of the UK, Europe, and North America, but many details of the life histories of the thousands of tropical butterfly species have not been documented.

REGIONAL ADAPTATIONS

Butterfly life histories depend on resource availability, including host plant availability and optimal temperatures. Consequently, different climatic regions and vegetation types may produce different phenologies in a single butterfly species.

For example, in western North America the Anise Swallowtail (*Papilio zelicaon*) is widespread, occurring in a variety of habitats, from mountaintops to the coast

and to desert areas. The phenology of this species differs greatly according to habitat. In an alpine habitat, the Anise Swallowtail completes just one generation a year, restricted by a short, three-month summer. In coastal areas of Oregon and California, it may produce two or three generations a year, taking advantage of nine to ten months of optimal temperatures for breeding and development. And in desert areas of the western USA, spring growth hosts the development of just one generation of Anise Swallowtails a year.

The phenology of alpine species is strongly controlled by temperature and the timing of snowmelt. The annual appearance of these butterflies may vary by two to four weeks from year to year, according to how much snow falls during the winter and how quickly it melts in spring. Most alpine butterflies overwinter as eggs or caterpillars, and need to undergo significant growth before they can pupate and emerge as butterflies.

Habitats characterized by long periods of favorable climate and food plant availability host butterflies with flexible phenologies. Generally, species in these habitats have continuous, overlapping generations. Ecosystems with these conditions are more characteristic of tropical and subtropical regions.

Voltinism:
the generation game

The number of generations or broods that a butterfly has in a year
is known as voltinism. Some species have one generation a year or
one every other year. Others may have two or three in a year, and
some tropical species that are not restricted by cold weather breed
continuously and may have six or more generations per year.

↙ The North American Orange Sulphur (*Colias eurytheme*), with many, overlapping generations developing in large alfalfa fields each year, may reach extremely high population density by the fall.

→ Green hairstreaks (*Callophrys* spp.) overwinter as pupae and have a protracted spring emergence period, spreading the risk of eclosing into unfavorable weather conditions.

THE ROLE OF CLIMATE

Voltinism is related to climate, which is governed by latitude and/or elevation. The better the weather, the more generations a butterfly can produce. All alpine butterflies have either a single generation per year, or may take two or three years to complete a generation. This is simply the result of the short window of opportunity available to them—they sometimes have only four to eight weeks for development and adult activity. Alpine species must cope with bad weather and unusually short or long seasons. Having evolved in this extreme habitat, they are naturally adaptable to changing and adverse conditions.

Multivoltine species can expand their populations with each successive generation, enabling an increasing abundance through a season. A good example is the Orange Sulphur (*Colias eurytheme*) in western North America, where populations can be enormous by the fall. However, late broods in multivoltine species are often at risk of failure due to the onset of early winter or drought. Producing multiple broods may be a gamble unless a significant proportion of the population takes a more conservative approach by terminating breeding well in advance of potential seasonal problems.

There can be issues for early species, too. An early spring can tempt some butterflies to eclose from their overwintering pupae, only to be grounded and perhaps even killed by a spring freeze. This has been observed in green hairstreaks (*Callophrys* spp.) that one year emerged in early April in 64°F (18°C) sunshine. After unseasonal heavy snowfall and temperatures between 21°F and 34°F (−6°C and 1°C) for four days, the entire population was wiped out. All was not lost for this green hairstreak population, though. Adaptation over millennia to unpredictable early spring environments has led this species to evolve an unusually lengthy period of eclosion. Happily, a few weeks later, newly eclosed green hairstreaks replaced the individuals killed by the unseasonal freeze.

NYMPHALIS ANTIOPA

Mourning Cloak

Living a long, lazy life

SCIENTIFIC NAME	*Nymphalis antiopa* (Linnaeus, 1758)
FAMILY	Nymphalidae
NOTABLE FEATURES	Large, dark chocolate butterfly bordered by a blue-flecked yellow band
WINGSPAN	3–3¼ in (75–83 mm)
HABITAT	Gardens, parks, riversides, and forest glades

The Mourning Cloak is a champion in the long-life stakes. The species, which outside North America is known by the alternative common names of Camberwell Beauty and Grand Surprise, routinely lives for 10 to 12 months. However, it spends a good seven months of that time "sleeping."

The Mourning Cloak is one of the longest-lived butterflies, sometimes living a full year as an adult. It is common for year-old faded and worn adults to fly together with their sparkling, freshly eclosed offspring. The black undersides of the sleeping Mourning Cloak's wings become a grand surprise indeed when they click open to reveal dark purple, blue, and yellow.

The species is one of the first butterflies seen in early spring, when it spends time basking on warm rocks and sipping nectar from willow catkins. The females lay eggs on willows in April, and the gregarious caterpillars complete their development in early June. The heat of summer induces newly eclosed Mourning Cloaks to become sleepy, and they pass the summer estivating in sheds or dark forests. They come out of their torpor briefly in September to feast on fall sap flows and fermenting fruit. Then, by October, they find and retreat to hibernation quarters, where they sleep until the promise of spring finds them and wakes them up.

IMMATURITY: THE WEAKEST LINK

Mourning Cloaks are rarely common butterflies, and seeing dozens in a single habitat is unusual. Despite their tenacity as adults, surviving adversity and living the longest life a butterfly can have, the growth of populations is constrained. This appears to be a consequence of the perils of immaturity. Eggs, caterpillars, and pupae appear to be targeted by a voracious and effective community of natural enemies, ranging from minute parasitic wasps to viral and bacterial pathogens. Mourning Cloak caterpillars are gregarious for the first half of their lives, and they appear to attract attention from insect and avian predators, as well as the aforementioned wasps and pathogens. However, caterpillar development is rapid, lasting just two weeks— and it needs to be, to mitigate the losses. A Mourning Cloak female can produce 250 eggs, of which just two or three are likely to survive to become adults.

→ The long-lived Mourning Cloak has a vulnerable developmental period when many natural enemies including wasps, flies, and disease wreak havoc on caterpillar and pupal survival.

HETERONYMPHA MEROPE

Common Brown

Common but special

SCIENTIFIC NAME	*Heteronympha merope* (Fabricius, 1775)
FAMILY	Nymphalidae
NOTABLE FEATURES	Orange with black markings and a blue-centered black eyespot on each forewing
WINGSPAN	2¼–2½ in (57–63 mm)
HABITAT	Grassy areas in parks and forests

As its name suggests, the Common Brown is commonly found in its favored grassy habitats in southern and eastern Australia. It is one of the few butterflies to undergo seasonal dormancy at different stages of life: in the winter as a caterpillar, and in the summer as an adult butterfly.

The Common Brown may be common, but only because it has developed successful ways to cope with the extreme Australian climate. The butterflies eclose in spring, sometimes in huge numbers, with the males emerging a few weeks before the females. By the time the females make their appearance, the males are eager to mate with them and frantically search for partners. They seek out newly eclosed females by flying close to the ground and inspecting every nook and cranny. Once mated, a female will spurn all future mating attempts by lying flat on the ground with wings closed, resembling a dead leaf.

A TALE OF TWO DORMANCIES

Once mating occurs, the shy females immediately seek shelter and dormancy. They find a cool, dark place, and here they stay until the Australian summer heat has passed. Having performed their reproductive duty, the males enjoy summer for as long as they can, but all have died by the time the females emerge from estivation in early fall.

In this season of grape harvest, female Common Browns can sometimes be found feeding on fallen or discarded fermenting grapes. Without males to harass them, they fly lazily through grassy areas, laying their eggs as they go. Young caterpillars either enter hibernation immediately or after a few weeks of feeding, taking the whole winter to complete development.

→ The Common Brown is an adaptable, long-lived Australian butterfly that is often found abundant in grasslands, parks, and gardens— particularly in the spring, when the males dominate, and in the fall, when only the females fly.

GONEPTERYX RHAMNI

Brimstone

The original butterfly

SCIENTIFIC NAME	*Gonepteryx rhamni* (Linnaeus, 1758)
FAMILY	Pieridae
NOTABLE FEATURES	Yellow to whitish green, leaf-shaped butterflies
WINGSPAN	2½–3 in (60–74 mm)
HABITAT	Forest, hedgerows, gardens, meadows, and parks

The large yellow Brimstone may be the species that gave us the word "butterfly." According to some, it was given the name "Butter-colored Fly" by the Anglo-Saxons, which gradually morphed into "butterfly."

Along with orange-tips (*Anthocharis* spp.), the Brimstone is a true harbinger of spring, bringing joy and color to the still-bleak countryside. Male Brimstones are bright yellow, fly rapidly, and demand attention. Both sexes hibernate, but it's the males that appear first, followed a few weeks later by the whitish-green females. The species produces a single generation, and adults live for up to a year, making them one of the longest-lived European butterflies.

Brimstones are masters of camouflage in all stages of their life. They have an unusual scalloped wing shape that resembles a leaf when the butterfly is roosting or hibernating in a bush.

The caterpillars are green, perfectly matching their buckthorn (*Rhamnus* spp.) host plants, and the pupae are shaped to mimic the appearance of a curled leaf.

GETTING FIT FOR SPRING

Brimstones eclose in late June or early July and spend the rest of the summer feeding on flowers, especially nectar-rich purple flowers such as thistles and teasels. Feeding intensifies as fall draws near and the butterflies build up their fat reserves for hibernation. They select a hibernation spot in September, invariably in a bush or among the leaves of ivy (*Hedera helix*), holly (*Ilex aquifolium*), or blackberry (*Rubus fruticosus*), where they look like just another leaf.

Drawing on their fat stores, the butterflies remain dormant until March, when the males take flight and replenish their energy reserves by feeding on spring flowers such as primroses, violets, and daisies. The males are spry and ready when the first females appear, most of which probably mate on their first early spring flight. Since the weather is often cool at this time of year, a mating pair of Brimstones can sometimes remain paired for many days.

Camouflaged pupa

The bulbous shape and green color of the Brimstone pupa helps it blend into the leafy environment in which it is formed.

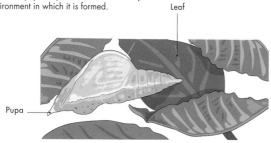

Leaf

Pupa

→ The Brimstone is a natural wonder of early spring, bringing a splash of yellow to the gray countryside, as it flies up and down country lanes.

THARSALEA HELLOIDES

Purplish Copper

Adaptable and pugnacious

SCIENTIFIC NAME	*Tharsalea helloides* (Boisduval, 1852)
FAMILY	Lycaenidae
NOTABLE FEATURES	Males are brown with a purple sheen, while females are orange and black-spotted
WINGSPAN	1–1¼ in (25–32 mm)
HABITAT	Almost everywhere, from vacant lots to wildlands

The Purplish Copper is a common and widespread species in western North America, and is adaptable in terms of habitat and seasonality. It can be found from early spring to late fall in habitats ranging from pristine to urban.

The Purplish Copper is a feisty little butterfly. Despite having a wingspan of just 1 in (25 mm), males defend their low perches by aggressively flying at intruders that are sometimes much larger than themselves. This tough approach is also manifested in the ability of Purplish Coppers to flourish in all kinds of habitats. They seem equally at home in weedy urban parking lots as they do in pristine mountain meadows. The species is one of a handful of butterflies that can occupy weedy and unkempt farm landscapes where pesticides are often used.

Being adaptable, ready to change, and ready to move on is the mantra of this butterfly. It is not a species to miss an opportunity to continue breeding and building its populations. This is shown by its habit of laying eggs late in the summer, some of which remain dormant and overwinter while the others take the chance to develop and produce another generation of caterpillars. If the gamble pays off, this second generation of adults fly in the fall. This aggressive life strategy ensures the species exploits resources for as long as they are available, yet has a "backup plan" (some non-developing, hibernating eggs) should the opportunism fail.

SLUGLIKE CATERPILLARS

Purplish Copper caterpillars feed on docks, sorrels, knotweeds, and smartweeds, which are rarely in short supply anywhere. Hatching caterpillars measure just 1.5 mm in length and grow to about ¾ in (20 mm) in six weeks. The sluglike green caterpillars are hard to find since they expertly match the color of the leaves they eat. They usually rest on the underside of leaves and feed by eating holes randomly through them, leaving a transparent membrane in their wake. Caterpillars pupate near the host plant on old leaves or debris. They vary greatly in color but usually match the background on which they are formed.

→ The Purplish Copper is a feisty and adaptable little butterfly equally at home in a parking lot as they are a pristine mountain meadow.

BICYCLUS ANYNANA

Squinting Bush Brown

Eyespots—now you see them,
now you don't

SCIENTIFIC NAME	*Bicyclus anynana* (Butler, 1879)
FAMILY	Nymphalidae
NOTABLE FEATURES	Decorated with eyespots on its fore- and hind wings
WINGSPAN	1⅓–1¾ in (35–45 mm)
HABITAT	Savannah and open forest, especially near the coast

Multicolored, bulls-eye patterns on butterfly wings— called eyespots for their resemblance to vertebrate eyes—are found throughout the brushfoots, but are most common among satyrs. These seemingly simple wing pattern elements of the Squinting Bush Brown take center stage in efforts to understand the relationship between genes, development, and an organism's physical features.

A butterfly's wings start to develop under its skin during the late caterpillar stage and they continue this process in the pupa. Early in development, an eyespot epicenter is formed between two wing veins. From this tiny spot, a chemical signal diffuses out in all directions, forming a gradient that is most concentrated near the center and becomes more dilute with distance. The surrounding cells detect the compound, and wing scales produce different colors depending on its concentration. Cells with the highest concentration develop a small circle of white scales. Then, after a concentration threshold is passed, the scales are black. Finally, in the outer fringes, where the signal is most dilute, the scales become golden-yellow. Beyond that, the scales are the normal background color.

SEASONAL POLYPHENISM

The undersides of the Squinting Bush Brown's wings are decorated with numerous eyespots—especially if the caterpillar develops during the Afrotropical rainy season. However, if the caterpillar develops in the dry season when water is scarce, the concentric circles of color will appear smaller or disappear altogether. This predictable change in wing patterns is known as "seasonal polyphenism," and the ambient temperature during the caterpillar stage determines whether the adult will have vibrant eyespots or not. Each form is thought to be best suited to the season in which it occurs. The lack of eyespots on the dry-season form makes it difficult for birds and other vertebrate predators to find them. However, attacks on the dry-season form by praying mantids are more often deadly because they attack the head. Larger eyespots in the wet-season form make the butterflies more conspicuous to birds and mantids. While this increases predation by vertebrates, it increases survival after a praying mantis attack because mantids target the eyespots on the margin of the hind wing. This is an expendable part of a butterfly's body, allowing it to fly away, alive, but missing an eyespot or two.

→ The small, faint eyespots of this female Squinting Bush Brown (left) indicate that she developed during the dry season, while her mate (right) went through his caterpillar stage in the wet season, as evidenced by his larger, more conspicuous eyespots.

DEFENSE &
NATURAL ENEMIES

Concealment and evasion

The first rule of defense is to hide. Butterflies, caterpillars, and pupae are very good at hiding. Where are all the butterflies on a cloudy day? Hiding. Where are all the caterpillars and pupae? Hiding. Where are all the enemies of butterflies? Seeking.

HIDING FROM THE ENEMY

Hiding from an enemy might be one of the oldest and most effective defense strategies used by potential prey. It is usually the first defense for butterfly eggs, caterpillars, pupae, and adults. The use of camouflage, subterfuge, and simple evasion can be successful.

Being active by night is an excellent way for caterpillars to avoid daytime enemies such as birds, and large predatory insects, including hunting wasps. There is still a risk of being found by nocturnal predators, but these are less common in most habitats.

During the day, many caterpillars hide under leaves, on the ground under rocks, or in some other hiding place. Most fritillary, satyr (Satyrinae), and skipper (Hesperiidae) caterpillars rest by day on or in the soil or

← Caterpillars of grass-feeding brushfoots like this Ochre Ringlet *(Coenonympha tullia)* blend in superbly with their surroundings.

↗ Caterpillars of some gossamer wings burrow into the stems, seedpods, or seeds of their host plants for refuge and food, such as this caterpillar of the Western Tailed Blue *(Cupido amyntula)*.

→ Caterpillars of many skippers create shelters on their host plants by silking leaf or grass edges together, such as this Mardon Skipper *(Polites mardon)* caterpillar.

at the base of their host plants. They are all colored in some shade of brown, gray, or black to blend in with the environment. When darkness falls, skipper and satyr butterfly caterpillars climb up their grass host plants to feed, avoiding the mice, shrews, and voles that may be hunting on the ground at night.

Another concealment strategy is to hide within a refuge. In some species such as blues and hairstreaks (Lycaenidae), which have small caterpillars, the host plant itself becomes a refuge—the caterpillar burrows into its stem, seedpods, or seeds, disappearing from the outside world, at least temporarily. Caterpillars of some skippers and duskywings (Hesperiidae) use silk produced from tiny spinnerets near their mouthparts to tie leaves or grass blades together to make a refuge. They often hide within this by day, coming out only to feed at night. Banana Skipper (*Erionota thrax*) caterpillars create tubes from banana leaves that they cover themselves with, hiding by day in the shelter.

BLENDING IN

Hiding from one's enemy without actually going anywhere is the illusion known as camouflage or crypsis. Blending in with your background is probably the most common strategy employed by eggs, caterpillars, and pupae, and by adult butterflies when at rest. Coloration of eggs, caterpillars, and pupae is often shaped by the need to be inconspicuous and not attract attention. Thus, they are often green-yellow to match their host plants. Grass-feeding caterpillars are green or brown with longitudinal stripes, which helps them blend into a sea of grass, and flower-feeding caterpillars like those of some blue butterflies (Lycaenidae) are brightly colored to match their favored blooms.

Because of their immobility, pupae almost universally rely on hiding and camouflage as strategies to avoid becoming a predator's meal. Those formed on host plants will be green or yellow, while those on inert surfaces will be shades of gray, brown, or black.

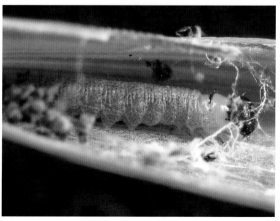

CLOSE YOUR WINGS AND DISAPPEAR

The lower surface of a butterfly's wings is usually less colorful than the upper surface. For example, the wing undersides of colorful brushfoot (Nymphalidae) butterflies such as Mourning Cloak (*Nymphalis antiopa*), tortoiseshells, Peacock (*Aglais io*), and commas (*Polygonia* spp.) are brown, gray, and black and do not attract attention. When these species roost in woody bushes and trees, they are perfectly camouflaged. Comma butterflies (*Polygonia c-album*) have scalloped, ragged-looking wings, breaking up their outline and making them look like a dead leaf when at rest. They also have striated markings, giving the lower wing surfaces the appearance of wood grain.

The white, yellow, and greenish butterflies in the family Pieridae generally have similar upper and lower wing surfaces, and these butterflies roost mostly in leafy bushes and trees. The wings of the yellow to whitish-green Brimstone (*Gonepteryx rhamni*; see pages 202–203) are scalloped and perfectly match the leaves of the ivy (*Hedera helix*) vines on which it frequently roosts.

DODGE AND ELUDE

If all the concealment and camouflage tactics fail, then evasion is another option. A sleeping butterfly that is disturbed will fly away. Most butterflies are almost impossible to catch by hand. Creeping up slowly and

← The ragged-looking and scalloped gray-black wings of the Oreas Anglewing (*Polygonia oreas*) allow it to hide in a dark forest habitat for overwintering.

←← The leaf-like shape and color of the Autumn Leaf (*Doleschallia bisaltide*) allow it to be camouflaged on the forest floor and masquerade as a dead leaf.

quietly to a butterfly that is busy feeding offers the best chance, but the odds are that it will fly off at the last minute. Brushfoots (Nymphalidae) and swallowtails (Papilionidae) in particular are hard to catch thanks to their fast, high, and erratic flight. Cabbage whites (*Pieris rapae* and *P. brassicae*) do not have speed, but they are very erratic in their flight behavior, which makes it hard for birds to follow and capture them.

In order to evade a predator, a disturbed caterpillar may fall to the ground to make itself difficult to find. Getting back to the plant again might be a problem, but some caterpillars get around this by suspending themselves on a silk lifeline, then winching themselves back up again when the threat passes.

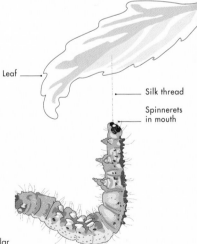

Leaf
Silk thread
Spinnerets in mouth

A lifeline made of silk

When a predator threatens and a caterpillar needs to escape, it may simply drop from its leaf perch. Falling to the ground could prevent its return to the host plant, so the caterpillar releases a silk thread from spinnerets in its mouth, from which it hangs in safety. When the threat has passed the caterpillar winches itself back up to the leaf.

Aposematism: giving a warning

Some butterflies, caterpillars, and pupae contain toxins and advertise this fact to potential enemies, a strategy called aposematism. The bold colors of the Monarch (*Danaus plexippus*) or Plain Tiger (*Danaus chrysippus*) are not just striking and beautiful, but also send a clear message to would-be predators: attack me at your peril!

WEAPONIZING CATERPILLAR CHEMISTRY

In addition to their warning coloration, swallowtail (Papilionidae) caterpillars (and the Caterpie Pokémon) have devised a way of presenting their defense chemicals to potential predators without the need for the predator to take a bite—which could, of course, be fatal. They do this via an eversible, Y-shaped gland behind the head that shoots out when the caterpillar is threatened. This bright orange or red gland, known as an osmeterium, glistens with a pungent secretion full of acids and terpenes that repels the would-be predator.

Caterpillars of fritillary butterflies (Nymphalidae) also have a defense gland similar to that of the   swallowtails, but positioned underneath the body instead of on top. In these species the gland secretes hydrocarbons and esters, including squalene, a known ant repellent.

USING PLANT CHEMICALS

Aposematism is the name given to the way butterflies and other animals advertise that they are not worth eating or attacking. Many butterflies defend themselves using chemicals. As caterpillars, they sequester noxious or poisonous chemicals from plants they feed on or make their own from benign chemicals. Consequently, caterpillars, pupae, and adults of many species are full of nasty chemicals designed to make birds and other vertebrates ill or even kill them.

Jane Van Zandt Brower and Lincoln Brower were American entomologists who discovered that Monarchs and their relatives contain poisons that could make scrub jays and blue jays "barf." The Browers' barfing jays was an early chapter in the field of "chemical ecology," from which we have learned much about how insects and other animals synthesize compounds or repurpose dietary toxins to defend themselves. Most butterflies that are distasteful or toxic to predators advertise their toxicity with bright colors and garish wing patterns.

LET MY COLORS BE YOUR WARNING

Caterpillars and butterflies that defend themselves with chemicals are boldly colored and patterned in an effort to make sure their toxicity is readily recognized and remembered by predators. Black, white, red, orange, and yellow in contrasting bands, stripes, and spots are the warnings most often used.

Milkweed butterflies in the genus *Danaus*, such as the Monarch and the Plain Tiger (*D. chrysippus*; see pages 248–249), all have similar black, orange, and white warning coloration and striped and spotted patterning as caterpillars and adults. The milkweeds these butterflies feed on as caterpillars contain cardiac glycosides, which they sequester to give them their toxicity.

← The caterpillar of the Monarch butterfly flaunts its toxicity via bold yellow, white, and black striping.

↑ The yellow-studded magenta caterpillars of the Atala (*Eumaeus atala*) butterfly warn predators of the toxic compounds ingested from its cycad host plant.

← The bold, distinctive colors of the Atala butterfly warn potential predators to stay away.

Checkerspots (*Euphydryas* spp.) feed as caterpillars on plants such as plantains and figworts that contain another group of toxic chemicals, iridoid glycosides. Like the cardiac glycoside-sequestering species, checkerspot caterpillars, pupae, and butterflies are warningly colored using white, black, red, and orange colors.

Monarch pupae are toxic but not aposematic. They are green and rely on camouflage for protection, developing in out-of-the-way places. In contrast, checkerspot pupae are aposematic and are often found in exposed positions. The pupae of Old World crow butterflies (*Euploea* spp.) seem to be filled with molten gold or silver, warning of the toxic compounds within.

Many other caterpillars and adult butterflies are colored and patterned in a presumed aposematic way but have not yet been confirmed as distasteful. Swallowtail butterfly caterpillars are often aposematically colored in black, white, and yellow, and likely contain toxins from their umbellifer host plants.

Mimicry: copying others

The butterfly world has its share of mimics. Posing as another butterfly species that is actively avoided by predators because of its distastefulness or toxicity is a true case of being a sheep in wolves' clothing.

Mimicry occurs in many different animals, including snakes, frogs, bumblebees, and dragonflies, and it can take many different forms. On one end of the spectrum, all of the species that resemble each other are toxic. These are known as Müllerian mimics. They share a common color pattern because birds and other predators need to sample at least one individual with a particular color pattern to learn that it is noxious. This could kill the butterfly. If each species has its own pattern, then every predator will try to

eat one of each. If multiple toxic species living in the same place look alike, then each predator only needs to eat one individual of one species to learn to avoid all of them. The cost is shared among all species. At the other end of the spectrum, palatable Batesian mimic species resemble truly noxious, aposematic "model" species. The model species are usually more common, and predators learn to avoid the color pattern by trying to eat the model species, which makes them avoid anything that looks like it.

In the tropics, things get even more complicated. Multiple species living in the same habitat can evolve the same color patterns: some of them are chemically defended Müllerian mimics and others are undefended Batesian mimics. Some of the mimetic species probably lie somewhere in between in terms of palatability. These "mimicry rings" don't just include butterflies. Day-flying moths, dragonflies, and damselflies can all join in, too.

MONARCH OR VICEROY?

The most famous butterfly example of mimicry is probably the Viceroy (*Limenitis archippus*), which mimics the Monarch (*Danaus plexippus*) in North America. It is remarkably similar to the Monarch, sometimes even fooling seasoned lepidopterists, but has an extra black vein on the hind wings and flies in a slightly different manner. In this textbook case, the Viceroy is not entirely a sheep in wolves' clothing, because its caterpillars accumulate their own toxins from feeding on willow host plants. So, it is doubly protected by pretending to contain cardiac glycosides from milkweeds and by actually having salicylic acid from willows.

TIGERS AND PALMFLIES

More than 50 species of palmflies (*Elymnias* spp.) can be found in tropical Africa, Asia, and Australasia. The vast majority of these palatable, palm-feeding butterflies seem to be undefended Batesian mimics of different chemically defended models. Several palmfly species mimic crow butterflies (*Euploea* spp.), while others mimic jezebels (*Delias* spp.), tigers (*Danaus* spp. and *Parantica* spp.), tree nymphs (*Idea* spp.), owls (*Taenaris* spp.), and, in Africa, *Acraea* butterflies. The palmflies

seem to have the ability to evolve almost any wing color or pattern to match the locally available toxic species. The Common Palmfly (*Elymnias hypermnestra*; see pages 178–179) has another trick. In some populations, males and females mimic the same, dark crow model species, while in other populations, the females (only) mimic orange tiger butterflies.

↑   In North America, the Viceroy butterfly (*Limenitis archippus*) is an accurate mimic of the Monarch butterfly. The horizontal band on the hind wings (that the Monarch does not have) separates the two.

←   Females of the Danaid Eggfly (*Hypolimnas misippus*) gain protection from predators by closely mimicking toxic and unpalatable relatives of the Monarch butterfly.

↑   The Pipevine Swallowtail
(*Battus philenor*) is distasteful to
predators because of toxic alkaloids
sequestered by the caterpillars when
feeding on pipevine host plants.

↗   The Spicebush Swallowtail
(*Papilio troilus*) is a successful mimic
of the distasteful Pipevine Swallowtail.
Experienced predators that avoid
eating toxic Pipevine Swallowtails also
avoid eating non-toxic Spicebush
Swallowtails.

Threats and scare tactics

Aggression itself can be a successful defense against aggression. Threatening and scaring your enemy can work if you are good at it and have surprise on your side, even if you do not have the resources or ability to go through with it.

KEEPING AN EYE OUT

When concealment, evasion, or hoodwinking your enemy fails, you can try threats and scare tactics. Unfortunately, a butterfly does not have many threatening devices it can employ. It does not possess any truly effective weapons as some insects have, such as the bombardier beetles' strategy of shooting a stream of toxic chemicals at a predator. To scare or threaten an enemy, a butterfly therefore needs the element of surprise.

→ False eyespots are part of the defense strategy of Western Tiger Swallowtail (*Papilio rutulus*) caterpillars. When threatened, the caterpillar swells up anteriorly, enlarging the eyespots.

↙ The Propertius Skipper (*Erynnis propertius*) caterpillar has a black head with two bright orange "eyes." In addition, this caterpillar will bare its mandibles if threatened.

→ When threatened, mature caterpillars of the California Sister (*Adelpha californica*) thrash their bodies wildly around, simultaneously displaying and moving their mouthparts as if to bite.

The classic scare tactic that some butterflies use is to show their "eyes"—not the tiny compound eyes on their heads, but the colorful, oversized eyespots on their wings. Having a few large eyespots can deter large vertebrate predators. If you disturb a resting Peacock (*Aglais io*) butterfly, for example, it will startle you by opening its wings suddenly and quickly, displaying four blue eyes on a bright red background. This visual display is accompanied by a hissing sound, which the butterfly makes by rubbing its forewings and hind wings. This may not seem scary to us, but it likely is to a bird, lizard, or mouse. Such a display of eyespots can indeed scare away potential vertebrate predators and, most importantly, protect the butterfly from harm. Some of the grass-feeding satyrs (Satyrinae) have large eyespots on their wings that may serve a similar scare function. Some of these also suddenly reveal these markings when disturbed.

JERKING AND THRASHING

Caterpillars have a greater number of threat and scare tricks up their sleeves than adult butterflies. Many display sudden movements when threatened, such as head-jerking and thrashing the front part of the body from side to side. A loud noise like a shout will sometimes evoke this response. This tactic is most effective when a large number of larvae do it in unison—as both European and North American tortoiseshell (Nymphalidae) caterpillars do when threatened.

When disturbed, California Sister (*Adelpha californica*) caterpillars thrash around and also bare their "fangs," displaying and moving their mouthparts, and caterpillars of some skipper (Hesperiidae) butterflies also display their mouthparts when threatened.

Caterpillars of tiger swallowtails (*Papilio* spp.) in North America have false eyes behind the head that they can puff up and enlarge when threatened, which in some species gives the appearance of a small snake.

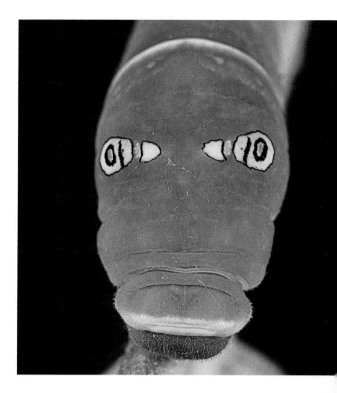

Diversions and deceptions

A little bit of deception can save your life. In the butterfly's "box of tricks" there are false heads, false eyes, and false antennae.

FALSE EYES AND FALSE HEADS

In addition to species such as the Peacock (*Aglais io*) that have big eyes on their wings, there are many others that have small eyes or dots on their wings, sometimes on the upperside, sometimes the lower side, and sometimes on both. These small fake eyes improve survival just like the big ones. However, instead of providing shock value, they deceive and deflect.

While having a few large eyespots can deter a predator, having many smaller eyespots deflect predator attacks to a non-essential part of the wings, as in the case of the Squinting Bush Brown (*Bicyclus anynana*; see pages 206–207). Small eyespots are usually positioned near wing edges, well away from the vulnerable butterfly body. Consequently, birds mistaking the spots for real eyes attack the more expendable outer parts of the wing, creating tears and cuts but allowing the butterfly to live another day. Proof of the success of this deception is seen in the fact that many butterflies have missing wing pieces where the spots once used to be.

Some hairstreak butterflies (Lycaenidae) take this deception one step further by carrying a false head on the tip of their hind wings, complete with a large "eye" and "antennae" in the form of little wing tails. When at rest, hairstreaks often rub their hind wings against the forewings with a curious circular motion. The tails usually twist in opposite directions, which causes them to move up and down independently. This adds to the appearance of an actual head, albeit at the opposite end of the real one. The deception really works; many hairstreaks are seen missing chunks of their hind wings.

← The long white-tipped hind wing tails of some Lycaenidae, including this Togarna Hairstreak (*Arawacus togarna*), are thought to mimic antennae, fooling predators into attacking expendable parts of the body, and allowing the butterfly to escape.

MASQUERADE

If they have no chemical defenses to advertise and cannot blend into the background, some butterflies resemble naturally occurring but inedible objects, a strategy known as "masquerade." Perhaps the most famous example is the Dead Leaf Butterfly (*Kallima inachus*; see pages 150–151). When sitting at rest with its wings closed, the unusual wings have the outline of a brown, dead leaf. A line from one end to the other is a dead ringer for a midrib—the big vein down the middle. Other markings resemble symmetrical cross-veins and asymmetrical patches of black fungus. Even within the same population, there is tremendous variability between Dead Leaf Butterflies, which helps prevent predators from developing a search image for them. Masquerade is not just employed as a tactic by adult butterflies; young caterpillars of many swallowtails, including the Old World Swallowtail (*Papilio machaon*), resemble bird droppings. During later caterpillar stages, once they've had more time to store up defensive compounds from their host plants in the parsley family (Apiaceae), the caterpillars are more conspicuously aposematic. Another common ruse is for pupae to resemble twigs, as in the Common Mime (*Papilio clytia*).

PLAYING DEAD

Unless they are scavengers, predators are not interested in eating dead meat and are instead looking for a live meal. Some butterflies and caterpillars try to take advantage of this preference by pretending to be dead. When picked up in cool conditions, the Mourning Cloak (*Nymphalis antiopa*) will hold its legs close to its body and remain motionless, even when placed on a surface. By refusing to stand or fly, this butterfly feigns death and most predators will move on. Many caterpillars will also feign death when disturbed, lying motionless until the threat has passed.

Aggregation:
safety in numbers

The mathematical biologist W. D. Hamilton (1936–2000) proposed his "selfish herd" theory in 1971. Its central tenet states that an animal seeks a central position in a group to reduce the danger to itself. Some butterflies and their caterpillars are great exponents and beneficiaries of this strategy.

KINGS OF AGGREGATION

Aggregation or gregariousness is a behavioral strategy that reduces the odds of any single individual being attacked or killed. Few sights match the spectacle of Monarch (*Danaus plexippus*) butterfly aggregations overwintering in Mexico and California. In Mexico, millions of butterflies gather each winter in a small area of high-elevation coniferous forest, adorning practically every inch of the trees here. The numbers in California are smaller, but the sight is still impressive. Gazing at a single tree clothed with hundreds of Monarchs is an unforgettable experience.

Smaller-scale aggregations are found in some other species, including the Common Crow (*Euploea corinna*) butterfly in Australia, and a few other brushfoots (Nymphalidae) in Africa. In Europe, small aggregations of overwintering Peacock (*Aglais io*) and Small Tortoiseshell (*A. urticae*; see pages 242–243) butterflies are often found in sheds and outbuildings.

There is safety in numbers for caterpillars and pupae, too. Like many moths, caterpillars of the Mexican Madrone (*Eucheira socialis*) (Pieridae) butterfly spin a silken shelter in which the larvae and pupae spend most of their time. Caterpillars and pupae of the ant-tended Australian Common Imperial Blue (*Jalmenus evagoras*; see pages 246–247) are also densely aggregated, perhaps because the right combination of host plant and attending ant is hard to find, concentrating them in a small area. These dark aggregations are so obvious at the tops of their wattle host plants that colonies of the butterfly can be spotted from a moving car.

↑ Least Skippers (*Ancyloxypha numitor*) congregate at a mud puddle in Canada.

← Snowberry Checkerspot (*Euphydryas colon*) butterflies drawing salt and other minerals from wet ground. Aggregations reduce the odds of any one butterfly being attacked, compared to solitary feeding.

Predators in the butterfly world

Butterflies are prey to an enormous number and variety of natural enemies, an umbrella term that includes vertebrate and invertebrate predators, parasitoids, and pathogens. Predators are the lions and tigers of the butterfly world, directly eating butterflies, eggs, caterpillars, and pupae.

← Reptiles and amphibians are important natural enemies of butterflies, particularly in tropical regions.

↗ Birds are likely the most important vertebrate predators of butterflies in all habitats and on all continents, and most butterfly defenses are aimed at these enemies.

AN ABUNDANCE OF ENEMIES

Eggs, caterpillars, pupae, and adult butterflies are all vulnerable to attack by natural enemies. The appearance, lifestyle, and behavior of all butterflies are shaped to a large extent by strategies and modifications to avoid being eaten. Crypsis, distastefulness, gregariousness, aggressiveness, and mimicry have all evolved as defenses against natural enemies.

From hundreds of eggs laid by every female butterfly, only a handful will reach adulthood, largely as a result of the depredations of natural enemies. Predators are one form of butterfly enemy, and seek, catch, and consume their prey. All butterfly stages fall prey to many types of vertebrate predators such as birds, mice, frogs, toads, lizards, and snakes, to a whole array of invertebrates ranging from spiders to praying mantids.

LARGE PREDATORS

The meandering flight of some butterflies, such as the satyrs (Satyrinae) and whites (Pieridae), has likely evolved to make it as difficult as possible for birds to catch them. Most successful bird attacks on butterflies probably occur early or late in the day, when conditions are cooler and butterflies are either resting or slow-moving. Birds are thought to be important as predators of eggs, caterpillars, and pupae, and likely take a substantial toll on these immature stages. Some birds specialize in feeding on butterfly eggs and caterpillars, particularly when provisioning their young.

Mice, voles, and shrews search relentlessly for food on or close to the ground, and caterpillars form a significant component of their diet. In spring, fritillary caterpillars live on the forest floor feeding on violets and likely are food for shrews and mice. Butterflies that roost near the ground, such as blues (Lycaenidae), may also serve as mouse food.

Butterflies in aggregations are not immune to attack and represent a substantial food resource to a wise predator. Although Monarchs (*Danaus plexippus*) are chemically defended by cardiac glycosides, some enterprising species of birds have learned that not all parts of these butterflies are toxic and/or distasteful. For example, black-backed orioles (*Icterus abeillei*) tear open overwintering Monarchs in Mexico, eating their insides while ignoring the cuticle where the poisons are stored. Predation by this bird species is estimated to kill millions of overwintering Monarchs in Mexico each year.

Reptiles and amphibians also devour their fair share of caterpillars and butterflies, should the insects wander too close to these sit-and-wait predators.

SMALL PREDATORS

The insect and spider predators of butterflies are many and varied, and overall likely have a much greater impact on butterfly populations than vertebrates. There are about ten groups or families of invertebrate predators that prey on the immature stages of butterflies, each containing dozens or hundreds of species in any particular location. However, only spiders, praying mantids, ambush bugs, predatory wasps, and perhaps dragonflies are common predators of adult butterflies.

PREDATORY SPIDERS

Spiders take a great toll on butterflies and caterpillars, and hunting and ambush spiders consume more than web-building spiders. Species of jumping, ambush, and crab spiders roam shrubs, bushes, and trees in search of prey. They are not fussy and will eat whatever insect they find and can cope with in terms of size. Crab spiders lurk in flowers, usually perfectly camouflaged, waiting for an unsuspecting visitor—often a butterfly. The spider throws its especially long front legs around its butterfly victim in a welcoming embrace, then inserts its fangs into the body and sucks the butterfly's life away. The population density of spiders is considerable in most habitats, so butterfly encounters with these predators must be frequent.

PREDATORY WASPS

Predatory wasps are important predators of butterflies in all life stages. Paper wasps and yellowjacket wasps are highly visible and aggressive predators. Not all species attack butterflies, but some do seek out eggs, caterpillars, and pupae. One species, the European paper wasp (*Polistes dominula*), exerts a high toll on butterfly populations in places such as New Zealand and North America, where it is an invasive species. This wasp establishes large populations in suburban areas and will remove every caterpillar in a neighborhood.

European paper wasp attacks on caterpillars are particularly gruesome. They surgically dismantle and remove all the parts of a caterpillar, even its alimentary canal. The wasps then take these back to their nests to

↖ Spiders that hunt by roaming plants, such as this jumping spider (Salticidae), capture and kill caterpillars and adult butterflies like this Boisduval's Blue (*Icaricia icarioides*) butterfly.

← The European paper wasp is an invasive species in North America, New Zealand, and Australia, where it may have a big impact on butterfly populations by killing large numbers of caterpillars, particularly in urban areas.

↑ An orchid mantis dines on a newly eclosed Common Palmfly (*Elymnias hypermnestra*) female before she has a chance to dry her wings.

MANTIDS, AMBUSH BUGS, AND DRAGONFLIES

Praying mantids lurk on or near flowers waiting for insects, including butterflies, to visit. When they do, they are swiftly grabbed by these true tigers of the insect world, which pull off their victims' wings and slowly chew through their bodies. Ambush bugs are smaller, cryptic versions of praying mantids, lurking in flowers ready to seize their prey. Sometimes the butterflies they capture are much larger than they are. Dragonflies catch their prey on the wing but are rarely observed doing so.

feed to their developing young. While Monarch butterfly pupae appear to be too tough for European paper wasps to pull apart, they have been observed waiting for a Monarch to begin eclosing and then attacking it. These wasps will even attack Monarch butterflies while they are feeding on flowers.

PREDATORY FLIES

Flies are not everyone's idea of a butterfly predator, but there are indeed a few families that attack and kill butterflies. The most visible and numerous of these are the large, fearsome-looking robber or assassin flies (Asilidae), which are not picky about what they attack and eat. They lie in wait until a suitable prey item flies past, then streak out to capture it in midair. Fast-flying and skilled in aerial pursuit and take-down, a robber fly will grab a butterfly in midair with its stout, spiny legs and bring it to the ground or a suitable perch. It then injects a fluid into its victim to paralyze it and make it easier to digest. After that, the fly selects a part of the butterfly body that is easy to insert its short, stout proboscis into—often the eye—and then spends the next hour or so sucking the juices from the hapless victim.

PREDATORS OF IMMATURE STAGES

Other insect and arachnid predators, ranging from mites and stink bugs to ants, prey on butterfly eggs, caterpillars, and pupae. Butterfly eggs are defenseless concentrated capsules of protein to tiny predators such as the mites and thrips that roam across plant surfaces. These diminutive predators suck out the contents of butterfly eggs, making them shrivel and disappear. Ants are also partial to feeding on butterfly eggs.

↖   Dragonflies catch their prey on the wing and butterflies are often taken, such as this Viceroy butterfly, being chewed by a large dragonfly.

←   Mites and thrips may feed on butterfly eggs. These eggs of a white butterfly (Pieridae) are being chewed by a thrip.

↗   Lady beetles are famed for feeding on and helping control aphids, but they will also consume butterfly eggs and small caterpillars if they come across them.

Other predators that wander over butterfly host plants include predatory beetles such as ladybird beetles. Although they more often feed on aphids, providing a valuable service for gardeners and farmers in the process, ladybird beetles will also attack and feed on butterfly eggs and small caterpillars. Ground beetles will take caterpillars from low-growing butterfly host plants such grasses and forbs, and therefore are likely to be important predators of brushfoot browns and fritillaries.

There are many kinds of true bugs that kill caterpillars, including shield bugs, stink bugs, and assassin bugs. These wander plants looking for prey, locating them by detecting and responding to the aromas produced by feeding caterpillars and sometimes by the noises they make while eating. True bugs feed by inserting their needlelike mouthparts into the fleshy body of a caterpillar and sucking up the juices. There are many species of tiny true bugs that also kill many butterfly eggs and small caterpillars. These species are just a few millimeters in length, so their impact on butterfly populations is hard to assess, but it may be substantial, particularly since they can vector pathogens.

Another predator that prefers to feed on aphids but will also feed on butterfly eggs and small caterpillars is the lacewing. The larvae of these net-winged green predators are voracious feeders that look like miniature alligators, complete with fearsome jaws.

Ants have varied relationships with butterflies. The majority of ant species see butterflies, their eggs, and caterpillars as a food source. Some ants will climb into bushes or trees and harvest all the eggs and caterpillars they find there. The caterpillars of many lycaenids (Lycaenidae) have a special gland from which they produce a sugary solution used to seduce ants into becoming bodyguards instead of predators. The Common Imperial Blue (*Jalmenus evagoras*; see pages 246–247) is probably the best-studied example of ant-tended butterfly larvae. Instead of attacking the caterpillar and taking it back to the nest, the ants instead gather around the caterpillar taking sips from the gland nectar. A pair of retractable scent glands surround this gland and seem to communicate warning signals to ants when in distress. Any other predator that approaches is rapidly turned away by the ants. Many metalmark caterpillars are also defended by

ants, but instead of one nectar gland they have two on the ends of paired tentacles.

Caterpillars and pupae that live with ants are also able to communicate with them acoustically. But since ants lack ears, the vibrations travel through the host plant or other surface and are detected with the ants' feet. Some species have taken ant symbiosis to the next level. Larvae of many carnivorous caterpillars manage to avoid being attacked by the ants that protect the aphids or other plant-sucking insects that they eat. Still other species, like the Moth Butterfly (*Liphyra brassolis*; see pages 94–95), live inside ant nests feeding on the ants themselves. Much like host plant associations, the relationships between caterpillars and ants is often fairly specific. Not all ants are acceptable partners.

Parasitoids: killing from the inside out

If predators are the lions and tigers of the butterfly world, then parasitoids are the aliens. Just like the extraterrestrial organism in the movie *Alien*, insect parasitoids develop within their caterpillar and pupal hosts, bursting forth from them once their development is complete.

THE UNSEEN KILLERS

Unlike parasites, which take resources from their hosts but do not normally kill them, parasitoids cause the death of their host. Insects that are parasitoids are important in agriculture and home gardens as biological control agents of herbivorous insects that are plant pests. While butterflies are not generally pests—apart from the infamous cabbage whites (*Pieris rapae* and *P. brassicae*) and fruit-feeding hairstreak butterflies—insect parasitoids are a big reason that most butterfly eggs, caterpillars, and pupae die before becoming an adult butterfly.

Parasitoids that target butterflies are generally small to minute wasps or medium-sized flies, and so are rarely seen. Parasitic wasps are nothing like the large, striped, aggressive predatory insects most people think of when the word "wasp" is mentioned. Instead, many are black or brown, and are difficult to see without a hand lens or microscope. Parasitic flies are usually larger than parasitic wasps and often look like house flies. However, they can be distinguished from the latter by their bristly rear end.

For most butterflies, parasitoids are as important as predators in regulating populations, if not more so. Affected butterfly eggs, caterpillars, and pupae often fail to hatch, develop, or eclose, instead producing single or multiple parasitoids that burst forth like the aforementioned alien.

← Cocoons of the parasitic wasp (*Cotesia glomerata*) after the larvae of the wasp has consumed the body contents of a Cabbage White (*Pieris brassicae*) caterpillar.

↗ This checkerspot (*Euphydryas* sp.) caterpillar has had its insides eaten by larvae of a parasitic wasp. The larvae are exiting the caterpillar and forming cocoons, from which a new generation of parasitic wasps will emerge.

TO EVERY BUTTERFLY, A PARASITOID

Butterfly eggs, caterpillars, and pupae are hosts to a great number of species of fly and wasp parasitoids. Many target only a single or closely related butterfly species, and most butterflies have parasitoids that are specific to them.

The female parasitoid lays her egg or eggs in or on a single host life stage—usually the egg, caterpillar, or pupa. The tiny larva then rapidly develops within this life stage, emerging as an adult either in the same host life stage or a subsequent stage. Wasps often inject their eggs into the host, whereas most parasitic flies lay their eggs on the caterpillar or pupal host and can be seen. When the tiny fly maggot hatches, it bores into the host. Some flies have a different strategy, laying thousands of eggs all over host plants, with caterpillars then ingesting the eggs as they feed.

A CLEVER MEAL PLAN

Once inside the host body, the parasitoid larva initially
starts feeding on non-essential parts, such as fat. It then
moves on to more important internal organs, ultimately
killing the host. The parasitized caterpillar may appear
quite normal until just before death, when its color may
change, it ceases feeding, and it becomes quiescent. The
parasitoid then breaks out through the host's skin.

Parasitoids can slow down the development of their
host to give them more time to complete their own
development. For example, some caterpillar parasitoids
do not exit their host until after it has become a pupa.
In other cases the fully fed parasitoid larva forms a cocoon
adjacent to the dead caterpillar, from which the adult wasp
or fly emerges after a few days.

Parasitoids can be a serious problem for communal
caterpillars. The development of parasitoids is rapid, and

↖ A caterpillar with a parasitic
tachinid fly egg on its head. A maggot
will hatch from the egg, bore into the
caterpillar and feed on internal organs.

↗ Aggregating caterpillars of the
California Tortoiseshell (*Nymphalis
californica*) may be heavily parasitized
by flies and wasps, causing dramatic
swings in the population of this species.

↗↗ *Sturmia bella*, a parasitic
tachinid fly, a recent arrival in
the UK from Europe, which may be
responsible for the recent population
decline of the Small Tortoiseshell in
the UK.

several generations can occur during the period when aggregated caterpillar hosts are available. This can result in substantial mortality to populations of these butterflies.

POPULATION REGULATORS

Population regulation by parasitoids is thought to be a major factor in the boom-and-bust cycles of some butterflies, including the California Tortoiseshell (*Nymphalis californica*) in western North America. In some years this species is extremely abundant, but a population boom is usually followed by a number of years when it is far less so. California Tortoiseshell caterpillars are gregarious on wild lilac (*Ceanothus* spp.) bushes and can defoliate large areas. Parasitoid populations likely build up within these aggregations, ultimately causing substantial mortality among their butterfly hosts.

A parasitoid is also thought to be behind the decline of one of Britain's most beloved butterflies, the Small Tortoiseshell (*Aglais urticae*). Fifty years ago, the Small Tortoiseshell was seemingly common everywhere— in the countryside, in towns and gardens, and all across the landscape. Today, however, its populations are much reduced, with only small numbers seen in a typical summer. In the late 1990s, the fly parasitoid *Sturmia bella* appeared in the UK, and research has shown that it is now the most common cause of death for caterpillars of the Small Tortoiseshell, killing almost half of them. Add this new mortality factor to all the other pressures butterfly populations have to face today, and it is easy to understand why the Small Tortoiseshell has suffered badly in the twenty-first century.

Parasites: taking resources but not killing

A true parasite is an animal that lives at the expense of another, while not killing it. It is not in the best interest of a parasite to kill its host. Two butterfly parasites have been studied extensively; one is a protozoan, the other a mite.

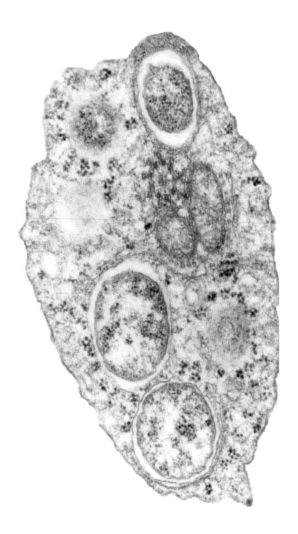

PROTOZOAN PACKS A PUNCH

Butterflies are rarely troubled by parasites that reduce viability or fitness. However, there are two exceptions that have been documented and there may be others that are yet to be discovered.

A protozoan with the tongue-twisting name of *Ophryocystis electroscirrha* (abbreviated OE) is a common parasite of the Monarch (*Danaus plexippus*) and a few closely related species. Dormant spores of this parasite are found on the cuticle of the Monarch, in and around its body, and on the wing scales. When female Monarchs oviposit, a few spores adhere to each egg, which the young caterpillar then eats upon hatching. Some spores are transferred to milkweed plants when butterflies are laying eggs or nectaring, and these may also be consumed by caterpillars.

Once inside the caterpillar, spores produce sporozoites, which reproduce in the gut during the life of the caterpillar and pupa. Just before eclosion, the parasite produces spores that then coat the newly forming butterfly's body. If there are thousands of spores, then this may cause the wings of the butterfly to be malformed, and in severe cases the butterfly may be unable to fly and will soon die. Under natural conditions this appears to be rare.

Monarch butterflies and OE conduct an ongoing and never-ending war to outcompete and outwit each other. Under natural conditions this leads to a relative balance between the two species, with neither gaining the upper hand. However, captive rearing of Monarchs for multiple generations can tip the balance toward OE, causing substantial mortality for the butterflies.

A COLORFUL AND BENIGN PARASITE

Many European butterfly species are parasitized by bright red mite larvae, which are very conspicuous on their hosts. The mite larvae lurk on grasses and flowers, and then attach themselves to butterflies when they visit or are resting on these plants. The mite larvae invariably attach themselves to the butterfly's thorax, and once there insert their mouthparts and suck up the butterfly's body fluids. Surprisingly, perhaps, mites feeding on a butterfly do not appear to have any effect on its life span or flight capabilities, and a single butterfly may have up to ten mites attached to it. As well as feeding on the butterfly, the mites benefit by using it for transport, enabling their dispersal over the landscape. Mites parasitize butterflies in other parts of the world too, particularly tropical regions. However, virtually nothing is known about these interactions, other than that they are unlikely to be harmful to butterfly lives, merely an annoyance.

SEX-CHANGING PARASITES

Undoubtedly the most common parasites of butterflies are bacteria. Numerous bacterial species have evolved the ability to infiltrate the bodies of host insects to be passed to the next generation in a female's egg cells.

↗ Red mite larvae are sometimes found attached to butterflies but they appear to do no harm and may just be using them as transport.

← The bacterium *Wolbachia pipientis* could infect up to half of all insects worldwide, including butterflies. In some populations, it changes the sex ratio to favor females.

These bacteria cannot be transmitted in sperm, so it is to the bacterium's advantage if there are more females than males in the population. More females equals more bacteria in the next generation. Thus, the parasites commandeer the butterfly's biology to overproduce females, and the mechanism they employ depends on the bacterial species and strain. *Wolbachia pipientis*, commonly called *Wolbachia*, infects between one-third and one-half of all insects on the planet. It can be benign, or it can favor the production of female offspring in several ways. Sometimes, it causes male embryos to die before they can become eggs. This prevents male caterpillars from competing with female caterpillars for food. Other kinds require that the mated male and female are infected with the same strain of *Wolbachia*, and still other varieties enable females to lay fertile eggs without sex (parthenogenesis). The mechanisms by which *Wolbachia* and other parasitic bacteria like *Spiroplasma* and *Cardinium* enter and then leave a population are poorly understood.

Pathogens: viruses, bacteria, and fungi

Like people, butterflies can catch diseases. Epidemics can wipe out local butterfly populations and are part of the natural ecology of some species. Butterfly diseases can be caused by viruses, bacteria, or fungi, and can strike at any stage of the insect's life cycle.

OVERCROWDING AND MOISTURE

Pathogens are infectious microorganisms that injure or kill their hosts. Bacteria, viruses, and fungi are the most common groups of pathogens. The diseases they cause can have a rapid and major impact on butterfly populations, primarily affecting eggs, caterpillars, and pupae. Caterpillar communal feeding and moist or humid conditions increase the chances that a pathogen will have a significant effect on the life of a butterfly.

COMMERCIAL EXPLOITATION

One naturally occurring bacterium, *Bacillus thuringiensis* (Bt), has been exploited as a caterpillar pest control agent. This pathogen is available commercially and used (often on a large scale) to control pest caterpillars like those of various conifer-feeding moths. Some Bt strains are effective against pestiferous caterpillars, while others target flies or beetles, leaving butterflies and moths unharmed. Some fungal pathogens have also been commercialized for use in pest insect control.

A VIRAL SOUP

Viruses are the most important and notable pathogens affecting caterpillars. Caterpillars dying from viruses become limp as their internal organs break down and liquefy. Consequently, infected caterpillars adopt characteristic hanging poses, attached by their front or rear legs and with fluid dripping from one or both ends.

Caterpillars of the Compton Tortoiseshell (*Nymphalis l-album*) in North America are particularly prone to a virus that can wipe out populations locally for a number of years. Proliferation of this virus is likely enhanced by the gregarious nature of the butterfly's caterpillars. Viruses may be transmitted by caterpillars feeding on contaminated leaves or passed on by infected females to their eggs.

BACTERIAL AND FUNGAL INFECTIONS

Bacterial diseases and fungi are probably relatively common in butterfly populations but rarely result in the kinds of widespread outbreaks that occur with viruses. Sublethal bacterial and fungal pathogens may have impacts on the longevity, fecundity, and general success of adults.

Fungi are the most poorly understood type of butterfly pathogen and occur mostly in tropical and subtropical regions where rainfall and relative humidity are high. Naturally occurring pathogens are the least understood natural enemies of butterflies. Undoubtedly, adult butterflies also get diseased, but we know virtually nothing about the pathogens affecting this life stage.

← A caterpillar of the Compton Tortoiseshell has a viral infection breaking down internal organs and making the body limp.

→ A caterpillar dying from toxin produced in its gut by spores of *Bacillus thuringiensis*. This bacterium is naturally occurring, and several strains are commercially available as a biopesticide.

Small Tortoiseshell

Familiar nettle-loving butterfly

SCIENTIFIC NAME	*Aglais urticae* (Linnaeus, 1758)
FAMILY	Nymphalidae
NOTABLE FEATURES	Vividly marked in reddish-orange, black, yellow, and blue
WINGSPAN	1¾–2½ in (44–63 mm)
HABITAT	Gardens, lanes, meadows, parks, and roadsides

The Small Tortoiseshell is one of Europe's most familiar butterflies. Bright, cheerful, and colorful, it is a habitat generalist that is at home in cities and on mountaintops. Keeping a patch of stinging nettles for the caterpillars and a butterfly bush for adults may entice a breeding colony to take up residence in a home garden.

These colorful butterflies are familiar throughout Europe because they are frequent inhabitants of urban parks and gardens as well as the wider countryside. If you have a butterfly bush, you will almost certainly find Small Tortoiseshells feeding from the flowers in the summer. In the UK, the species is not as common today as it once was, likely due to a fly parasitoid that kills the caterpillars. The stinging nettle is the only host plant of Small Tortoiseshells. The caterpillars and their silken nests are often seen on these plants, which are unpopular with people because of their stinging hairs. The abundance and ubiquity of stinging nettles is an important part of why Small Tortoiseshells can be seen anywhere on the landscape.

YOUNG, SPINY, AND GREGARIOUS

Small Tortoiseshell females lay untidy clumps of up to 100 eggs, which may take her as long as an hour. They usually choose a large leaf near the top of a nettle plant in a sunny location. After a couple of weeks, the tiny caterpillars eat their way out of the eggs and begin spinning a communal web. The gregarious caterpillars live within this web, which protects them from predators. As they grow, they move across the nettle plants, constructing new webs. Then, halfway through their caterpillar lives, they abandon web-building and begin to live apart. At this point, their defense changes to dependence on their spines, and their ability to jerk their bodies from side to side, spout green fluid, and fall to the ground if threatened. Mature caterpillars will also sometimes construct leaf tents in which to hide from predators.

→   The Small Tortoiseshell is one of Europe's prettiest and most familiar butterflies, well adapted to living near people but may be found anywhere on the landscape.

PAPILIO BAIRDII

Oregon Swallowtail

Oregon state insect

SCIENTIFIC NAME	*Papilio bairdii* (Edwards, 1866)
FAMILY	Papilionidae
NOTABLE FEATURES	Yellow, blue, and black butterfly with a pair of red spots
WINGSPAN	3½–4 in (89–102 mm)
HABITAT	Canyons, riversides, and slopes in arid lands

The Oregon Swallowtail occurs in the Pacific Northwest of North America, and was recently separated as a distinct species from the Old World Swallowtail (*Papilio machaon*), which it closely resembles. Although very similar to its Old World cousin, which occurs in Eurasia, Canada, and Alaska, the Oregon Swallowtail has unique host plants and habitats.

In the arid interior of North America's Pacific Northwest, the Oregon Swallowtail is a butterfly of basalt canyons, river banks, and slopes. This is a very specialized habitat that differs markedly from its sister species, the Old World Swallowtail (*Papilio machaon*), which is a habitat generalist. Instead of relying on the umbellifers (celeries and parsleys) favored as caterpillar host plants by Old World Swallowtails, the Oregon Swallowtail uses wild tarragon (*Artemisia dracunculus*), in the daisy plant family (Asteraceae). The life of this butterfly is closely tied to its single host plant, which remains green and palatable during summer in hot, arid environments. This enables the butterfly to produce two generations each year.

MASQUERADING AS FECES

Young Oregon Swallowtail caterpillars behave like other swallowtail caterpillars to protect themselves: they pretend to be bird excreta. This exploits the sensible notion that potential avian predators of juicy caterpillars have no desire to investigate their own feces as a possible item of nourishment. The disguise is simplicity itself: a dark brown to black background with a well-placed white "saddle." But to a bird, the caterpillar looks like dung and is treated as such, leaving it to feed and grow.

Of course, most bird excreta are not large or three-dimensional, so as the caterpillar grows, the disguise wears thin. When the caterpillar molts for the third time, it unveils a new defense: an aposematic pattern. This is bold and flashy, and, most importantly, bears a warning that says: "I am not good to eat." From masquerading as bird excreta, the Oregon Swallowtail is transformed into a striking caterpillar with black and yellow spots and bands. And just in case any bird does not get the message and gets too close, the caterpillar will dramatically pop out a pungent forked "snake tongue" (osmeterium) from behind its head, as a final deterrent.

→ The Oregon Swallowtail is the New World name for this gorgeous butterfly that is also the state insect of Oregon, USA.

JALMENUS EVAGORAS

Common Imperial Blue

Adapted to living with ants

SCIENTIFIC NAME	*Jalmenus evagoras* (Donovan, 1805)
FAMILY	Lycaenidae
NOTABLE FEATURES	Turquoise blue upperside and buff underside with black markings and hind-wing tails
WINGSPAN	1¼–1½ in (32–37 mm)
HABITAT	Open scrubland or subtropical eucalypt forest

Without ants to protect them, caterpillars and pupae of the Common Imperial Blue don't stand a chance against their enemies, so they have many adaptations to ensure a healthy entourage of their meat ant guardians (*Iridomyrmex* spp.).

Common Imperial Blue females only lay eggs if they can detect the right species of ants on their wattle (*Acacia* spp.) hostplants, and they seem to prefer laying clumps of eggs along ant trails on the trunks of saplings. The caterpillars are highly attractive to ants. Their skin is dotted with many small glands that secrete amino acids, and there is a single, large gland near the hind end (the dorsal nectary organ, DNO) that secretes a solution of sugars and more amino acids. All of these nutrient-rich secretions come at a cost. Caterpillars raised without ants in the laboratory result in much larger butterflies, and larger butterflies are more successful at mating and laying eggs. However, caterpillars deprived of ants in nature are killed by predators and parasitoids nearly 100 percent of the time. The nutritional supplements provided by the caterpillars are highly beneficial for the ants; colonies with access to caterpillars grow larger.

ASSOCIATION WITH ANTS

When disturbed, a pair of powderpuff-like organs are extruded from pockets near the DNO. Their presence drives the ants wild, perhaps because they smell like the ants' alarm pheromone. When attacked by a parasitoid or predator, the smell of these organs puts the ants on high alert. If the number of ant guardians is low, the caterpillars and pupae can vibrate at specific frequencies that resonate through the woody stems of their host plant to beckon more ants. The caterpillars and pupae are frequently found in groups of around a dozen or more individuals, often covered by a writhing layer of ants. Adult butterflies are rarely encountered far from ant-covered wattles capable of nurturing young caterpillars. Rather than flying around to look for mates, males sit around ant-covered pupae until a female ecloses. A mad scramble ensues, and one male— usually the largest—mates with her before her wings have had a chance to harden.

→ A female Common Imperial Blue rests on a wattle branch defoliated by caterpillars of the same species.

DANAUS CHRYSIPPUS

Plain Tiger

Old World cousin of the Monarch

SCIENTIFIC NAME	*Danaus chrysippus* (Linnaeus, 1758)
FAMILY	Nymphalidae
NOTABLE FEATURES	Orange wings with black-and-white markings and a white-spotted black body
WINGSPAN	2¾–3 in (73–79 mm)
HABITAT	Arid open areas, gardens, and parks

The Plain Tiger is an uninspiring name for this attractive African-Asian butterfly, which has prompted the widespread use of more majestic monikers such as the African Queen. This butterfly, a relative of the Monarch (*Danaus plexippus*), also uses chemistry, unpalatability, and warning coloration for protection.

For a long time, the Plain Tiger was thought to occur in Australia, and was known there as the Lesser Wanderer, being a smaller cousin of the Wanderer, which is the Monarch's common name in Australia. However, recent research has shown that the Lesser Wanderer is actually a different species, *Danaus petilia*. The Plain Tiger, like its Australian cousin, is definitely a wanderer, roaming African and Asian landscapes in search of its milkweed host plants. It prefers open arid areas, but can be encountered in just about any landscape or habitat.

WHAT YOU EAT WILL PROTECT YOU

Like Monarch caterpillars, Plain Tiger caterpillars feed on milkweeds and gain protection for themselves and the subsequent adults by sequestering the cardiac glycosides that these plants contain. Both the caterpillars and adult butterflies are aposematic, with striking black, white, orange, and red colors, warning all would-be vertebrate predators that they do not taste good. In many cases, though, this is a bluff, because not all Plain Tigers contain poisons. This is because many milkweeds contain little or no cardiac glycosides. It is not only non-poisonous Plain Tigers that take advantage of their poisonous cousins—other completely palatable butterflies mimic the distinctive coloration and patterning, hoping to fool predators into believing they are distasteful. In addition, it is likely that male Plain Tigers add to their unpalatability as adults by feeding from flowers that contain pyrrolizidine alkaloids. The males also use these chemicals to synthesize pheromones that are necessary to attract females for mating.

→ The Plain Tiger is a close relative of the Monarch and is often confused with it, leading to the alternative common name of African Monarch.

MIMACRAEA MARSHALLI

Marshall's Acraea Mimic

A colorful algae-eating mimic of the
Plain Tiger

SCIENTIFIC NAME	*Mimacraea marshalli* (Trimen, 1898)
FAMILY	Lycaenidae
NOTABLE FEATURES	Mimic of the Plain Tiger and its co-mimics
WINGSPAN	1⅓–1½ in (35–40 mm)
HABITAT	Open forest from 3937–5577 ft (1200–1700 m) in elevation; often perched on tree trunks

Things are not always what they seem. Marshall's Acraea Mimic resembles co-occurring toxic butterflies, but its larval diet of lichen and algae render it perfectly edible to its enemies.

As its genus name *Mimacraea* implies, most members of this group are Batesian mimics of the butterfly genus *Acraea*, a diverse group of unpalatable butterflies found mostly in Africa. Marshall's Acraea Mimic strongly resembles toxic Pierre's Acraea (*Acraea encedana*) and Common Acraea (*Acraea encedon*). However, these butterflies are Müllerian mimics of an unrelated toxic species: the Plain Tiger (*Danaus chrysippus*). When found together, they comprise a "mimicry ring" of different but similar-looking species, some of which are defended by poisons that deter predators.

With around 235 species, the genus *Acraea* is one of the most diverse in the world. Each species is believed to be inedible to predators, and they advertise this with a unique combination of black, red, orange, white, or yellow markings. Remarkably, all *Mimacraea* species resemble a distantly related *Acraea* or *Danaus* (tiger) model species. Research on the genomes of Müllerian mimetic butterflies from the

Neotropics has revealed that extreme similarity between species is often due to the exchange of wing patterning genes when closely related species interbreed. However, interbreeding between an Acraea Mimic and its *Acraea* model would never be successful because the two are too distantly related. The uncanny similarity between Marshall's Acraea Mimic and the Plain Tiger is therefore due to evolutionary convergence.

ALGAE-FEEDING BUTTERFLIES?

Marshall's Acraea Mimic belongs to the tribe Liptenini, and all members of this group have an unusual caterpillar diet in that they consume only algae and lichen. Although various moths can consume these ubiquitous foodstuffs, within butterflies, this behavior seems to be restricted to this wholly African group. The adults have unusual diets, too. Instead of feeding on flowers, they suck the honeydew of scale insects—the sweet, sticky excretion left over from their diet of plant sap.

→ A Marshall's Acraea Mimic hangs
upside-down while at rest.

THREATS &
CONSERVATION

Habitat loss and fragmentation

Becoming homeless through habitat loss is the primary reason butterflies are less common today than they were 50 years ago. A butterfly without a home will not survive to reproduce, and the population will be extirpated.

URBANIZATION AND SPRAWL

Urban expansion drives butterfly habitat loss in the Global North, although agricultural expansion is the main reason for habitat loss in the Global South. Concomitantly, butterfly populations are declining at a greater rate in urban and near-urban areas than in the wider countryside. Today, urban areas are expanding twice as fast as their human populations. In the next

decade alone, urban land cover is likely to increase by more than 400,000 square miles (1 million sq km). The Xerces Blue (*Glaucopsyche xerces*) is the only butterfly acknowledged to have become extinct in North America as a result of urbanization. The last individual of this species was seen in the early 1940s. This butterfly lived on the San Francisco peninsula in coastal sand dunes that were bulldozed for development. The nearest alternative habitat was many miles away, out of reach of the isolated peninsular population. While the Xerces Blue is a telling example of what can happen when a limited habitat is destroyed, there are thousands more butterfly species around the world with populations that have greatly diminished as a direct result of habitat reduction and fragmentation through urban development.

← The last remaining Xerces Blue butterfly was seen in its San Francisco dune habitat in the early 1940s. This butterfly is the best-known example of a species being made extinct by destruction of its habitat.

↗ Modern, broad-acre agriculture has been responsible for the destruction and fragmentation of many butterfly habitats over the past 50 years, particularly in Europe and North America.

AGRICULTURAL EXPANSION

The expansion of commercial agriculture has led to the destruction of many natural habitats all over the world. Together with urban sprawl, agricultural development is a major continuing threat to butterfly populations, and has been responsible for the decline of many species.

Modern broad-acre cropping systems are worse for butterfly populations than small-acre family holdings, which are usually interspersed with natural or semi-natural habitats. More than 100 million almost contiguous acres (40 million hectares) of the American Midwest are dominated by wheat, corn, and soybeans. Little natural butterfly habitat exists in this region now, but before corporate agriculture the landscape was varied, more species-rich, and supported a more diverse butterfly fauna. European agriculture has also expanded and intensified, drastically reducing the availability of natural habitats. Clearing tropical forests for oil palm, rubber, sugar cane, cattle pastures, and other agricultural commodities has devastated the diversity and abundance of butterflies and other biodiversity.

In South America, deforestation for timber and livestock cause unfathomable areas of rainforest to be cut down each year. In Asia, forests are cleared primarily for growing oil palm, but other tree crops, such as rubber, *Acacia*, and *Eucalyptus*, also contribute to the problem.

EUROPEAN BUTTERFLY DECLINE

Butterfly observation and study has a longer history in Europe than almost anywhere else. Thus, the decline in butterfly diversity and abundance in the face of urbanization and agricultural intensification in this region has been better documented than elsewhere. In most European countries, 30–80 percent of butterfly species have declined in just the last few decades and 8–30 percent no longer exist in some locations.

In 2022, scientists from Butterfly Conservation in the UK compiled a Red List of the conservation status of Britain's butterflies. Of the 58 species on the list, 24 are categorized as threatened and only 29 are considered not to be of conservation concern. Since the 1950s, Britain has lost 97 percent of its flowering meadows and 80 percent of chalk grasslands. Habitat losses of such magnitude cannot help but reduce butterfly populations.

↖ There are around 240 species in the Oakblue genus *Arhopala*, and a quarter of these have historically been recorded in the Philippines, including this Sylhet Oakblue (*A. silhetensis*). After a half-century of intensive deforestation, less than a quarter of the islands' forests remain and many of these forest-loving species haven't been seen in decades.

↗ The Duke of Burgundy is restricted to habitats where its sole host plant, cowslip, grows, and is vulnerable when these habitats are fragmented or degraded.

BREAKING UP THE HABITAT

While not as serious as losing your home entirely, breaking up a habitat into smaller pieces can cause significant problems for butterflies. A general rule is that a smaller habitat is not as good as a larger habitat. This is because it will contain fewer resources and space, and will therefore limit the number of butterflies that can live there. Smaller populations are less able to cope with disaster or adapt to changing conditions. The key for many butterflies as to whether they can cope with fragmentation of their habitat by roads, farms, or buildings is the distance to the next area of suitable habitat. Can they fly there easily? Easy linkage between small pieces of habitat can sometimes enable

as many butterflies to exist as would do so in a single large habitat.

Wide-ranging mobile butterflies that use a variety of host plants for their caterpillars, such as the Painted Lady (*Vanessa cardui*, see pages 174–175), are well adapted to small patches of habitat, and so fragmentation has the least impact on them. In contrast, butterflies restricted to a single or few host plants, such as the Duke of Burgundy (*Hamearis lucina*, see pages 274–275), and those that are unable to fly far, can be badly impacted by habitat fragmentation.

Climate change: stressful extremes

Climate change and the extremes in weather that come with it are threats to butterfly survival. As global temperatures increase, there will be winners and losers. The winners will be those that can adapt and exploit changing conditions, while the losers will be those that are unable to change quickly enough.

CANARIES IN THE COALMINE

Butterflies are environmental indicators. When something goes wrong with their environment or habitat, they are among the first organisms to be affected and respond. Temperature guides butterfly life, and conditions that are too hot or too cold can be a problem. A warming climate may provide opportunities for enhanced breeding and range expansion in some species, but in others it may literally push them off a cliff.

Butterflies that may win in a warming world will be those that benefit from a few extra degrees of heat. This might allow three broods in a season instead of two, which may increase population size and sustainability. Warmer temperatures will also allow the northward range expansion of some species.

↗ The Melissa Arctic whose home habitat exists on mountain summits, faces an uncertain future as temperatures rise.

← The Silver-spotted Skipper has expanded its populations in the UK as warmer summers have created better opportunities for breeding.

In the UK, populations of the Silver-spotted Skipper (*Hesperia comma*, see pages 268–269) have increased since the turn of the twenty-first century. From fewer than 70 occupied sites, this butterfly can now be found in more than 200 locations, and the warming climate has played an important role in this resurgence.

Butterflies that may lose in a warming world are those that are adapted to cool or moderate climates and cannot readily or rapidly adapt to increasing temperatures. The classic example are high-elevation montane butterfly species. One such species is the Melissa Arctic (*Oeneis melissa*), which lives on mountain summits from Japan to Siberia and North America. What will happen to this butterfly when temperatures increase at the summit? It cannot move to cooler temperatures at higher elevations. Can it adapt to warmer temperatures or will it be pushed to extinction? Only time will tell.

EXTREME WEATHER AND POOR HABITATS

Extreme weather events such as heatwaves, flooding, unseasonal freezes, and droughts can have a major impact on butterfly populations. Long-term data collected on British butterfly abundance indicate a sustained effect of heatwaves on populations that may last for years. While more mobile older caterpillars and adult butterflies can seek shady, cooler locations, butterfly eggs and young caterpillars are particularly vulnerable to desiccation in extreme heat.

Fragmented or degraded habitats may offer less buffering against weather extremes than large, intact, high-quality habitats. For example, the deep, dark woods within a largely open butterfly habitat may provide a valuable cool refuge for butterflies when the weather gets too hot. Logging has decimated the oyamel fir forests where monarch butterflies take their winter refuge. Destruction of forests surrounding these small areas used by the butterflies has eliminated a climate buffer, making the overwintering masses vulnerable to lethal freezing events.

Pesticides: death, instant or slow

Butterflies are rarely the target of pesticides, but they are certainly vulnerable to them. Being in the wrong place at the wrong time can mean instant death. Today's pesticides are safer for people but still lethal to insects. However, just like the first-generation pesticide DDT, we are learning that today's pesticides can be slow, silent, and sustained agents of death for innocent species such as butterflies.

A NECESSARY EVIL?

Pesticides are a part of modern life. There are currently 8 billion humans on Earth, and everyone needs to eat. Without pesticides to prevent crop losses to insect pests, many more people would go hungry than already do. Fungicides kill fungal diseases on plants, herbicides kill weeds, and insecticides kill insects. However, synthetic chemicals such as these have the potential to affect all life in obvious or subtle ways.

A NEW SILENT SPRING?

Rachel Carson's book *Silent Spring* (1962) warned of the environmental harm caused by pesticides, and in particular the "silent" harm caused by the organochlorine insecticide DDT. This chemical is persistent in the food chain, causing subtle long-term problems for wildlife.

In the 1990s a new class of insecticides was introduced, which are now the most widely and commonly used insecticides in the world. These are the neonicotinoids, effective against pests. Like the organochlorines, however, they are persistent and cause sublethal impacts such as reduced reproduction and lifespans in butterflies and other non-target insects.

Caterpillars exposed to trace amounts of neonicotinoids show retarded and less successful development in the few butterfly species that have been tested. One study tested the effect of tiny doses of a neonicotinoid on the length of adult life of feeding adult Monarch (*Danaus plexippus*) butterflies. These Monarchs suffered 80 percent mortality over three weeks compared to 20 percent in butterflies not exposed to the neonicotinoid.

The European Union took notice of the science concerning detrimental effects of neonicotinoids on pollinating and beneficial insects, and implemented a ban on the outdoor use of these chemicals in 2018.

ORGANIC IS BEST

With so many unknowns on the impact of pesticides on butterflies, it is best to have an organic-only policy. Whether the habitat concerned is a garden or natural area, no effort should be spared to ensure it is pesticide-free to protect the butterflies and other beneficial insects living there from harm. By going organic in your garden, you will also avoid any other possible long-term impacts on your own health.

↗ With conventional agriculture comes pesticides, most of which are harmful to butterfly lives. Strides are being made to improve the interface between agriculture and beneficial insects like butterflies. Organic crop production, which greatly limits or removes the use of synthetic chemicals, is usually compatible with insect life.

→ Forest butterflies such as hairstreaks can suffer when insecticides are applied aerially to control moth caterpillar pests.

Farming and gardening for butterflies

The future for butterflies in countries with dense human populations or those dominated by agriculture will depend on how gardening and farming practices can be adapted to provide the resources that butterflies need for their lives.

FARMING WITH BUTTERFLIES

Farming and crop production today, and increasingly so in the future, has the potential to be butterfly-friendly. However, decades of pesticide use have taken a toll on the abundance and diversity of insects all over the world. Today, some farmers are transitioning to using chemicals that are "softer," meaning they are better targeted at pest species and spare other insects. The path to a pesticide-free future will have hiccups along the way, such as the unforeseen sublethal impacts of neonicotinoid insecticides (see pages 260–261). However, in some areas and with some crops, there is real potential to sustain viable butterfly habitats alongside crops.

↗ Blue butterflies (Lycaenidae) in the genus *Icaricia* are among the species that can be encouraged to live and breed in naturescaped vineyards in the Pacific Northwest of North America.

← "Beauty with Benefits" is the concept of utilizing native flowering plants in and around croplands to enhance populations of beneficial insects to improve biological control of pests and pollination.

BEAUTY WITH BENEFITS

In Washington state in the USA, wine grape production uses little pesticide. There are few pests and most are controlled biologically by native natural enemies. Biological control is further enhanced by restoring native plants and native habitat to areas around the vineyards, and sometimes even within the rows of grapevines. In addition to minimizing harm to butterflies and other pollinators from pesticides, this more natural approach also provides them with more habitat and resources. As a result, thriving populations of butterflies are now a feature of habitat-restored or "naturescaped" vineyards in Washington.

As agriculture shifts to biological control rather than chemical control for pest management, these benefits will be seen in other areas. Such opportunities are already being exploited in Australia and New Zealand, and are now beginning to appear in Europe.

↑ Grazing by wild or domestic animals can be important for sustaining breeding populations of butterflies such as the Silver-bordered Fritillary. Without grazing, the violet host plants of this species are shaded out and decline.

→ A wide variety of flowers that bloom throughout the summer and fall will attract a variety of butterfly species.

GRAZING CAN BE GOOD FOR BUTTERFLIES

Many butterflies that live in savannahs depend on grazing by wild animals to maintain their habitat. Without these grazers, grass growth would inhibit the survival of herbaceous butterfly host plants. When seedlings of woody plants are not removed by vertebrate grazing, bushes and shrubs may then grow, turning a meadow habitat into scrubland, which then excludes meadow butterflies.

The same process may occur on land grazed by farm livestock, including cattle, sheep, and horses. Such grazing may allow the presence of butterfly host plants that support meadow butterflies such as the Silver-bordered Fritillary (*Boloria selene*). Grazing also ensures ground-loving butterflies such as skippers (Hesperiidae) receive enough warmth and light for breeding and survival.

GARDENING FOR BUTTERFLIES

The idea of creating a butterfly garden has been around for some time— the British statesman Winston Churchill created one just after the Second World War. However, it is only in the past decade or two that it has really taken hold in the public imagination. Developing home gardens as "pocket nature reserves" is now recognized as an important strategy to help alleviate the decline in butterfly populations in urban areas. By creating a butterfly garden, you can do something meaningful for the conservation of butterflies, pollinators, and the other beneficial insects in your neighborhood.

NECTAR

To be successful, a butterfly garden must have the things that butterflies need. The number-one requirement is, of course, nourishment—primarily the nectar from flowers. Not all flowers are good suppliers of nectar, and butterflies have their favorites. Flowers that have flat heads, such as daisies, allow easy landing and perching for butterflies and so are popular with many species.

GARDENING FOR BUTTERFLIES

Creating a butterfly garden is an idea that has
been around since the Victorian era. However,
it is only in the past decade or two that the idea
has really taken hold. Developing home gardens
as "pocket nature reserves" is now recognized
as an important strategy to help us alleviate the
decline in butterfly populations in urban areas.
A butterfly garden can help conserve butterflies
and other beneficial pollinators.

Michaelmas daisies and other asters are also a major draw for many butterflies. Care should be taken when using Butterfly bush (*Buddleja* spp.) and Lantana, as these plants can be invasive and damaging to native plant communities in some areas.

As we saw on pages 66–71, butterflies also use other sources of nourishment, such as overripe fruit, wet mud, animal feces, and carrion. While animal carcasses and waste are probably not desirable in the home garden, providing a patch of wet ground where butterflies can sip and obtain extra salt and minerals is a good idea. A bowl of overripe fruit will sometimes also attract butterflies.

HOSTING CATERPILLARS

While attractive flowers will lure butterflies to a garden, they will leave once they have topped up with nectar. The second requirement is the host plant or plants that they need to lay their eggs on. Gardens with these plants might attract butterflies that will lay their eggs to become a nursery for the next generation.

For example, a patch of stinging nettles in an out-of-the-way area of your garden can provide a caterpillar food resource for a number of butterfly species in Europe, North America, North Africa, Asia, and Australia. Red Admirals (*Vanessa atalanta*), Small Tortoiseshells (*Aglais urticae*), and Peacocks (*Aglais io*) all use nettles, and will readily come to suburban gardens that feature these plants.

A milkweed patch in a suburban garden in North America or eastern Australia will often attract a female Monarch (*Danaus plexippus*) to lay eggs. Garden weeds, including docks in Europe and mallows in North America, are used by the caterpillars of coppers (Lycaenidae) and skippers (Hesperiidae), respectively. If you have a large garden and can set aside an area of native grass, it may be used by grass-feeding species such as browns (Nymphalidae) and skippers.

STRUCTURE

Butterflies prefer a habitat with structure and diversity. Having trees and bushes that they can perch in, open ground where they can bask, and corridors that they can patrol along gives them the ability to do the important tasks in their lives.

The larger the garden space, the more potential there is for attracting and sustaining a wider range of butterflies. However, very small gardens and even potted plants can still provide important "refueling stations" for passing butterflies. Creating a successful butterfly garden requires learning about the butterflies and plants specific to an area. A good knowledge of the butterflies likely to visit your garden and their nectar and host plant requirements is essential to ensure you create a local and effective butterfly oasis.

← Red Admiral females will lay their eggs and raise a family on out-of-the-way stinging nettle patches in suburban gardens and parks in Europe and North America.

→ The stunning Peacock is a frequent garden visitor in Europe, attracted by a range of butterfly-attractive flowering plants.

HESPERIA COMMA

Silver-spotted Skipper

Common Branded Skipper

SCIENTIFIC NAME	*Hesperia comma* (Linnaeus, 1758)
FAMILY	Hesperiidae
NOTABLE FEATURES	Silver-spotted golden skipper
WINGSPAN	1–1¼ in (25–32 mm)
HABITAT	Warm chalk grasslands, especially south-facing slopes

The Silver-spotted Skipper is the European name for a species that has a Holarctic distribution from North America to Asia. In North America, it is known as the Common Branded Skipper because a different North American species (*Epargyreus clarus*) goes by the European monicker.

This is a warmth-loving butterfly that reaches its UK northern limit in southern England. Here, it is restricted to warm, closely grazed chalk downland sites where its caterpillar food plant, sheep's fescue (*Festuca ovina*), grows. Both sexes spend a lot of their time basking or feeding on a wide variety of flowers.

The Silver-spotted Skipper needs a temperature of at least 68°F (20°C) to fly. When seeking a mate, the male perches on a grass blade in the sun and dashes out to investigate any passing insect. Females lay their eggs singly on sheep's fescue, preferring short plants (½–1½ in/1–4 cm tall), usually surrounded by bare ground. Females lay eggs only when ground temperatures are 77°F (25°C) or higher. The eggs overwinter and the caterpillars develop during spring, producing butterflies in July.

CLIMATE CHANGE WINNER

The Silver-spotted Skipper has benefited from a warming climate, and there are now estimated to be four times as many populations of this butterfly in southern England than there were 50 years ago. Another factor enhancing population growth in this species has been an increase in grazing of its grassland habitats by rabbits, opening up the warm microhabitats that the butterfly needs for basking and egg-laying. Conservation efforts based on providing warm microhabitats through careful and timely grazing by introduced livestock have also been part of the species' success story.

→ The Silver-spotted Skipper is among a small group of UK butterflies that are doing much better as summers become warmer and sunnier.

Wall Brown

Losing ground with climate change

SCIENTIFIC NAME	*Lasiommata megera* (Linnaeus, 1767)
FAMILY	Nymphalidae
NOTABLE FEATURES	Orange and black butterfly with up to ten eyespots on the forewings
WINGSPAN	1½–2¼ in (38–55 mm)
HABITAT	Grasslands with areas of bare and stony ground, including sand dunes and quarries

The Wall Brown used to be a common butterfly in the UK but has disappeared from many of its former inland haunts, likely because of climate warming. It has now become a predominantly coastal butterfly.

As its common name suggests, this butterfly is fond of basking on walls—or, indeed, any warm surface. The Wall Brown has a characteristic behavior of resting with its wings two-thirds open, reflecting warmth back onto its body to raise its temperature and allow a speedy getaway should it be needed. Today, Wall Browns in the UK occur in small, discrete populations on sand dunes and coastal cliffs. The females lay eggs low down on grasses, which are host plants of the caterpillars.

Wall Brown males are very territorial, resting in sunny spots, ready to dart out and intercept any passing butterfly. However, they also use other strategies for finding a female such as patrolling up and down a path, or hill-topping. Once a female is found, she is pursued until she lands on the ground where mating takes place.

CLIMATE CHANGE LOSER

The reason for the decline of the Wall Brown in the UK was a mystery until research suggested it had become a victim of climate change. Normally, the Wall Brown has two generations of adults, one in early May and the other in late July, with the half-grown caterpillars from the second generation overwintering. However, in warm summers— as experienced in recent years in the UK, especially in inland areas—they continue developing, resulting in a third generation of adults in September. Unfortunately, the weather in fall does not usually allow caterpillars from this third generation to reach the overwintering stage, meaning that they and the local population are extirpated. In cooler coastal areas the butterfly is not coerced into attempting an extra generation, which is why the coast now hosts most of the Wall Brown populations in the UK.

→ The Wall Brown is among a large group of UK butterflies that are experiencing declining populations. In this species, it appears that climate warming is having a negative impact.

ORNITHOPTERA RICHMONDIA

Richmond Birdwing

Large and spectacular Australian butterfly

SCIENTIFIC NAME	*Ornithoptera richmondia* (Linnaeus, 1767)
FAMILY	Papilionidae
NOTABLE FEATURES	Black with brilliant iridescent green markings
WINGSPAN	5–6 in (130–150 mm)
HABITAT	Lowland and upland subtropical rainforests

The spectacular Richmond Birdwing is the largest butterfly in eastern Australia. Unfortunately, it is threatened and populations have contracted and declined by more than half, mostly because of habitat destruction and fragmentation.

The Richmond Birdwing lives in lowland subtropical rainforests, where its host plants—native vines in the family Aristolochiaceae—grow. The butterflies are strong fliers, and mostly active high in the rainforest canopy in the early morning and near dusk. Courtship is elaborate, with the male chasing after the female and then hovering to sprinkle her with a pheromone. The caterpillars are large and bulky, and eat a huge quantity of food during their four to six weeks of development. There are two to three generations a year, with the pupae overwintering.

Richmond Birdwings are sexually dimorphic, meaning the sexes look different. The larger females have none of the iridescent green of the males, being dark brown or black with cream and yellowish markings.

A TOXIC DECEIVER

An introduced plant species in the same family as the butterfly's native host plants tricks female Richmond Birdwings into laying their eggs on it. However, this plant— Dutchman's pipe (*Aristolochia littoralis*)—is toxic to the caterpillars and none survive to reproduce.

The botanic lookalike is an important driver in the decline of the Richmond Birdwing and considerable effort is being made to remove as many of the vines as possible. Happily, a Richmond Birdwing recovery program is helping establish corridors between existing populations, as well as maintaining populations by planting and nurturing the native host plants.

Richmond Birdwing caterpillar

The Richmond Birdwing caterpillar lives solitarily and is prone to cannibalistic tendencies should more than one caterpillar occur on a single vine.

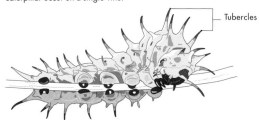

Tubercles

→ In 1870, the Richmond Birdwing occurred in the streets of Brisbane. Today, habitat loss and fragmentation means this butterfly can now only be found in two areas with rainforest remnants, north and south of Brisbane.

HAMEARIS LUCINA

Duke of Burgundy

Diminutive metalmark butterfly

SCIENTIFIC NAME	*Hamearis lucina* (Linnaeus, 1758)
FAMILY	Riodinidae
NOTABLE FEATURES	Brown and orange in a fritillary-like checkerboard pattern
WINGSPAN	1–1 ¼ in (25–33 mm)
HABITAT	Clearings in open forests, and grazed chalk and limestone grasslands

The Duke of Burgundy is the sole European representative of the metalmark family (Riodinidae). It is found in scattered colonies across most of the continent and in parts of temperate Asia.

The easiest way to tell the sexes of this butterfly apart is to count the legs: females have six legs and males have only four functional legs. The Duke of Burgundy was once widespread across much of England and Wales, but today there are just a few populations, primarily in southern England. The adults emerge in late April or early May and the species is single-brooded. The butterflies are most active on sunny mornings and rarely move far from their breeding sites. Both sexes bask in sheltered locations in grass or on small bushes. They prefer north- or west-facing slopes where the host plants, cowslips (*Primula veris*), tend to show more profuse growth.

IMPROVING STATUS QUO

In England, this butterfly has declined with the cessation of forest management practices that ensured maintenance of clearings, essential for survival of colonies. An estimated 98 percent of forest colonies disappeared between 1950 and 1990, and consequently the majority of extant Duke of Burgundy populations in England are found on chalk and limestone grasslands. This species is also vulnerable here, needing sufficient grazing by rabbits to ensure optimal growth of the cowslips and caterpillars.

However, the butterfly's status quo has improved in the past decade because of implementation of appropriate habitat management strategies by conservation organizations. Among these is the need to carefully manage grazing to ensure minimal damage to host plants and caterpillars.

Speckled pupa
The pupa of the Duke of Burgundy is light-colored, speckled, and unusually hairy for a pupa. Formed on or near the ground, the pupa overwinters.

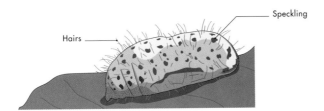

Speckling

Hairs

→ The Duke of Burgundy was once called the "Duke of Burgundy Fritillary" since its wings resemble those of true fritillaries in color and patterning.

PARALUCIA PYRODISCUS LUCIDA

Eltham Copper
Formerly extinct Australian butterfly

SCIENTIFIC NAME	*Paralucia pyrodiscus lucida* (Crosby, 1951)
FAMILY	Lycaenidae
NOTABLE FEATURES	Brownish-black with bright copper markings
WINGSPAN	1–1 1/16 in (25–27 mm)
HABITAT	Mixed eucalypt forest and open forest

The Eltham Copper is a subspecies of the Fiery Copper (*Paralucia pyrodiscus*), endemic to Australia. While Fiery Coppers can be found widely across eastern Australia, the Eltham Copper is restricted to just 25 colonies distributed between three regions in the southern state of Victoria.

The Eltham Copper was thought to have become extinct in the 1950s but was rediscovered at Eltham, a suburb of Melbourne, in 1986. Since then, much conservation interest has been invested in protecting this butterfly, which is categorized as endangered. In 2020, a newly discovered population of the Eltham Copper resulted in abandonment of plans to build hundreds of feet of new railway tracks in a Melbourne suburb. Loss and fragmentation of habitat is thought to be responsible for the decline of this butterfly.

ANTS AS BODYGUARDS AND GUIDES

The Eltham Copper is dependent on sweet bursaria (*Bursaria spinosa*), the sole host plant for caterpillar development. Another essential component to its butterfly is the presence of ants in the genus *Notoncus*. Eltham Copper caterpillars spend their days living in the underground nests of these ants. Every night when it is warm enough, the ants guide the caterpillars out of the nest and into the sweet bursaria to feed. The ants are always attentive and protect the caterpillars from predators. As a reward, the caterpillars provide sugary snacks for the ants from special glands. When dawn approaches, the ants guide the caterpillars back to the safety of the nest. This relationship continues for 10 or 11 months, until the caterpillars turn into pupae near or within the ant nest.

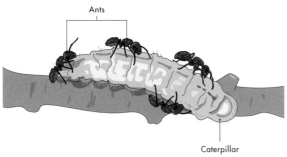

Ants

Caterpillar

Ants and caterpillars
Eltham Copper caterpillars owe their lives to ants! Ants protect and "farm" the caterpillars of this species. Their reward? A sweet sugary solution.

→ The Eltham Copper was once considered extinct, but is now a poster child for butterfly conservation in Australia. The status of this butterfly is now improving after attracting interest and community support.

GLOSSARY

All butterfly wings, like this wing of a Blue Morpho, are covered with thousands of tiny scales that overlap. The remarkable iridescent blue of a Morpho wing is generated by the physical structure of individual scales, rather than pigments which are responsible for most other colors on butterfly wings.

abiotic factors Non-living components of an ecosystem such as water, temperature, sunlight, nutrients, and the soil, which may affect the lives of butterflies.

aposematic coloration Warning coloration or patterns indicating unpalatability to potential predators.

chitin Polysaccharide forming the major constituent in the exoskeleton of insects.

cremaster Tiny hooks found at the terminal end of butterfly pupae enabling attachment and support.

crypsis The ability of an organism to conceal itself from natural enemies by having a color, pattern, or shape that allows it to blend into the surrounding environment.

diapause A period of physiologically enforced dormancy between periods of activity.

diurnal Active only during daylight hours.

eclose/eclosion The emergence of an adult butterfly from its pupa.

estivate To pass the summer in a state of dormancy or torpor.

extirpate The regional loss of a species.

frass Excrement produced by insects.

gregarious Butterflies or caterpillars in a cluster or a colony.

hemolymph Insect blood.

hibernal Winter dormancy.

Holarctic Northern faunal region including Europe, Asia, and North America.

hyperabundant In great abundance.

instar Different caterpillar stages demarcated by molting.

integument Sclerotized cuticle covering an insect's body.

lek/lekking Aggregation of males at a site to attract females or where females occur.

lepidopterist A person who studies butterflies and moths.

melanic Black form of caterpillar or butterfly.

morphology Form of an organism or its features.

neonicotinoids Class of persistent neuro-active insecticides chemically similar to nicotine.

New World Western hemisphere of the world, primarily the Americas.

Old World Eastern hemisphere continents of Europe, Asia, and Africa.

osmeterium Defensive, Y-shaped gland behind the head of late instar swallowtail caterpillars.

outbreaks A sudden and large increase in population.

ovariole Tubular component of the insect ovary.

oviposition Egg-laying.

ovipositor Egg-laying tube extension of the insect oviduct.

parasitoid A parasite whose parasitism results in the death of the host.

phenology The timing of biological events, often related to weather, seasons, and climate change.

proboscis Coiled feeding tube used by butterflies to feed on flower nectar.

pupa (pl. pupae) The resting stage when a caterpillar undergoes metamorphosis into an adult butterfly. Also called a chrysalis.

refugium (pl. refugia) A portion of a species' former range that it inhabits during a period of habitat loss.

riparian Moist habitat adjacent to watercourses or bodies of water.

sclerotized Hardened insect integument caused by deposition of sclerotin.

senesce Decline in vigor and capacity following maturity.

sequester Storing defensive chemicals ingested from host plants.

spermatophore A capsule produced by male butterflies containing sperm and nutrients.

sphragis A male structure that detaches during copulation and seals the female's mating duct.

venation The system of veins on an insect wing.

voltinism The number of generations in a year.

BUTTERFLY FAMILIES

Taxonomy, which concerns the names and classification of organisms, is constantly changing. Researchers describe new species and decide that some species should be merged into others on an almost daily basis. Genetic studies that infer evolutionary history are similarly changing views of what constitutes a genus, family, or subfamily. Recent work on the evolutionary history of butterflies by David J. Lohman and colleagues (Kawahara *et al.* 2023) suggests that numerous changes to butterfly classification are needed, and are not reflected in these tallies. This list summarizes the current (2023) taxonomic organization of butterflies, which are all classified in the superfamily Papilionoidea.

SUPERFAMILY PAPILIONOIDEA

7 Families

43 subfamilies

1,900+ genera

19,500+ species

FAMILY HEDYLIDAE

no subfamilies

1 genus

36 species

FAMILY HESPERIIDAE

13 subfamilies

640+ genera

4,200 species

Barcinae 2 genera

Chamundinae 1 genus

Coeliadinae 9 genera

Eudaminae 56 genera

Euschemoninae 1 genus

Hesperiinae 361 genera

Heteropterinae 15 genera

Katreinae 2 genera

Malazinae 1 genus

Pyrginae 113 genera

Pyrrhopyginae 35 genera

Tagiadinae 26 genera

Trapezitinae 19 genera

FAMILY LYCAENIDAE

7 subfamilies

450+ genera

5,500+ species

Aphnaeinae 17 genera

Curetinae 1 genus

Lycaeninae 9 genera

Miletinae 14 genera

Polyommatinae+Theclinae★ 353 genera

Poritiinae 58 genera

FAMILY NYMPHALIDAE

13 subfamilies

540+ genera

6,300 species

Apaturinae 19 genera

Biblidinae 40 genera

Calinaginae 1 genus

Charaxinae 18 genera

Cyrestinae 3 genera

Danainae 58 genera

Heliconiinae 33 genera

Libytheinae 2 genera

Limenitidinae 45 genera

Nymphalinae 56 genera

Pseudergolinae 4 genera

Satyrinae 270 genera

FAMILY PAPILIONIDAE

3 subfamilies

30+ genera

600 species

Baroniinae 1 genus

Papilioninae 20 genera

Parnassiinae 10 genera

FAMILY PIERIDAE

4 subfamilies

80+ genera

1,100+ species

Coliadinae 18 genera

Dismorphiinae 7 genera

Pierinae 61 genera

Pseudopontiinae 1 genus

FAMILY RIODINIDAE

3 subfamilies

140+ genera

1,500 species

Euselasiinae 10 genera

Nemeobiinae 17 genera

Riodininae 121 genera

★ *It has been demonstrated that Polyommatinae is nested within Theclinae, but the two subfamilies have not yet been formally merged.*

→ Orange Gull
(*Cepora aspasia*, family Pieridae)

RESOURCES

BOOKS

Ackery, P.R. and R.I.Vane-Wright. *Milkweed Butterflies, Their Cladistics and Biology.* (British Museum, 1984)

Agrawal, A.A. *Monarchs and Milkweed: A Migrating Butterfly, a Poisonous Plant, and Their Remarkable Story of Coevolution.* (Princeton University Press, 2017)

Aoki, T., S. Yamaguchi, and Y. Uemura. *Butterflies of the South East Asian Islands, Volume 3: Satyridae, Amathusiidae & Libytheidae.* (Plapac Co. Ltd, 1982)

Bascombe, M.J., G. Johnston, and F.S. Bascombe. *The Butterflies of Hong Kong.* (Academic Press, 1999)

Braby, M.E. *Butterflies of Australia: Their Identification, Biology and Distribution.* (CSIRO Publishing, 2000)

Brown, F.M. and B. Heineman. *Jamaica and Its Butterflies.* (E.W. Classey Limited, 1972)

Carson, R.L. *Silent Spring.* (Mariner Books Classics, 2022)

Clark, G.C., and C.G.C. Dickson. *Life Histories of the South African Lycaenid Butterflies.* (Purnell, 1971)

Corbet, A.S., H.M. Pendlebury, G.M. v.d. Poorten, and N.E. v.d. Poorten. *The Butterflies of the Malay Peninsula.* 5th Edition. (Malaysian Nature Society, 2020)

DeVries, P.J. *The Butterflies of Costa Rica and Their Natural History, Volume 1: Papilionidae, Pieridae, Nymphalidae.* (Princeton University Press, 1987)

DeVries, P.J. *The Butterflies of Costa Rica and Their Natural History, Volume 2: Riodinidae.* (Princeton University Press, 1997)

Eeles, P. *Life Cycles of British and Irish Butterflies.* (Pisces Publications, 2019)

Gallard, J.-Y. *Les Riodinidae de Guyane.* (Jean-Yves Gallard, 2017)

Hall, J.P.W.A. *Monograph of the Nymphidiina (Lepidoptera: Riodinidae: Nymphidiini) Phylogeny, Taxonomy, Biology, and Biogeography.* (Entomological Society of Washington, 2018)

Igarashi, S. and H. Fukuda. *The Life Histories of Asian Butterflies, Volume 1.* (Tokai University Press, 1997)

Igarashi, S. and H. Fukuda. *The Life Histories of Asian Butterflies, Volume 2.* (Tokai University Press, 2000)

James, D.G., Ed. *The Book of Caterpillars.* (The University of Chicago Press, 2017)

Kim, S.-S. and Y.-H. Ho. *Life Histories of Korean Butterflies.* (Sagyejŏl, 2012)

Larsen, T.B. *Butterflies of West Africa.* (Apollo Books, 2005)

Neild, A.F.E. *The Butterflies of Venezuela. Part 1: Nymphalidae I (Limenitidinae, Apaturinae, Charaxinae).* (Meridian Publications, 1996)

Neild, A.F.E. *The Butterflies of Venezuela. Part 2: Nymphalidae II (Acraeinae, Libytheinae, Nymphalinae, Ithomiinae, Morphinae).* (Meridian Publications, 2008)

Newland, D.E. *Discover Butterflies in Britain.* (WILDGuides Ltd, 2006)

Parsons, M. *The Butterflies of Papua New Guinea: Their Systematics and Biology.* (Academic Press, 1998)

Pennington, K.M., E.L.L. Pringle, G.A. Henning, and J.B. Ball. *Pennington's Butterflies of Southern Africa.* 2nd Edition. (Struik Winchester, 1994)

Pyle, R.M. *Chasing Monarchs: Migrating with the Butterflies of Passage.* (Houghton-Mifflin Company, 1999)

Pyle, R.M. and C.C. LaBar. *Butterflies of the Pacific Northwest.* (Timber Press, 2018)

Quicke, D.L.J. *Mimicry, Crypsis, Masquerade and Other Adaptive Resemblances.* (Wiley Blackwell, 2017)

Ruxton, G.D., W.L. Allen, T.N. Sherratt, and M.P. Speed. *Avoiding Attack: The Evolutionary Ecology of Crypsis, Warning Signals, and Mimicry.* 2nd Edition. (Oxford University Press, 2018)

Samways, M.J. *Insect Conservation: A Global Synthesis.* (CABI, 2020)

Scott, J.A. *The Butterflies of North America: A Natural History and Field Guide.* (Stanford University Press, 1986)

Shirōzu, T. *The Butterflies of Japan in Color.* (Gakushū Kenkyūsha, 2006)

Tennent, J. *Butterflies of the Solomon Islands: Systematics and Biogeography.* (Storm Entomological Publications, 2002)

Tsukada, E. *Butterflies of the South East Asian Islands, Volume 4. Nymphalidae (1).* (Plapac Co. Ltd., 1985)

Tsukada, E. *Butterflies of the South East Asian Islands, Volume 5: Nymphalidae (2).* (Plapac Co. Ltd., 1991)

Tsukada, E. and Y. Nishiyama. *Butterflies of the South East Asian Islands, Volume 1: Papilionidae.* (Plapac Co. Ltd., 1982)

Williams, W. *The Language of Butterflies.* (Simon & Schuster, 2020)

Willmott, K.R. *The Genus Adelpha: Its Systematics, Biology and Biogeography (Lepidoptera: Nymphalidae: Limenitidini).* (Scientific Publishers, 2003)

Xerces Society. *Gardening for Butterflies.* (Timber Press, 2016)

Yata, O. and K. Morishita. *Butterflies of the South East Asian Islands, Volume 2: Pieridae and Danaidae.* (Plapac Co. Ltd., 1985)

SCIENTIFIC JOURNAL ARTICLES

Cusser, S., N.M. Haddad, and S. Jha. "Unexpected functional complementarity from non-bee pollinators enhances cotton yield." *Agriculture, Ecosystems & Environment* 314: 107415 (2021)

Forister, M., C.A. Halsch, C.C. Nice, J.A. Fordyce, and T.E. Dilts. "Fewer butterflies seen by community scientists across the warming and drying landscapes of the American west." *Science* 371: 1042–1045 (2021)

Gilburn, A.S., N. Bunnefield, J.M.V. Wilson, and M.S. Botham. "Are neonicotinoid insecticides driving declines of widespread butterflies?" *PeerJ* 3 e1402 (2015)

James, D.G., L. Seymour, and T.S. James. "Population biology and behavior of the imperiled *Philotiella leona* (Lycaenidae) in south central Oregon." *Journal of the Lepidopterists Society* 68: 264–273 (2014)

James, D.G., L. Seymour, G. Lauby, and K. Buckley. "Beauty with benefits: Butterfly conservation in Washington State, USA, wine grape vineyards." *Journal of Insect Conservation* 19: 341–348 (2015)

James, D.G. "Imidacloprid at a rate found in nectar reduces longevity but not oogenesis in Monarch butterflies, *Danaus plexippus* (L.). (Lepidoptera: Nymphalidae)." *Insects* 10: 276 (2019)

Kawahara, A.Y., C. Storer, A.P.S. Carvalho, D.M. Plotkin, F.L. Condamine, M.P. Braga ... and D.J. Lohman. "A global phylogeny of butterflies reveals their evolutionary history, ancestral hosts and biogeographic origins." *Nature Ecology & Evolution* 7: 903-913 (2023)

Lowe, T., R.J. Garwood, T.J. Simonsen, R.S. Bradley, and P.J. Withers. "Metamorphosis revealed: Time-lapse three-dimensional imaging inside a living chrysalis." *Journal of the Royal Society Interface* 10: 20130304 (2013)

Reppert, S.M. and J.C. de Roode. "Demystifying monarch migration." *Current Biology* 17: R1009–R1022 (2018)

Rodder, D., T. Schmitt, P. Gros, W. Ulrich, and J.C. Habel. "Climate change drives mountain butterflies towards the summits." *Scientific Reports* 11: 14382 (2021)

Ruttenberg, D.M., N.W. VanKuren, S. Nallu, S.-H. Yen, D. Peggie, D.J. Lohman, and M.R. Kronforst. "The evolution and genetics of sexually dimorphic 'dual' mimicry in the butterfly *Elymnias hypermnestra*." *Proceedings of the Royal Society B* 288: 20202192 (2021)

Talavera, G. and R. Vila. "Discovery of mass migration and breeding of the Painted Lady butterfly (*Vanessa cardui*) in the sub-Sahara: The Europe-Africa migration revisited." *Biological Journal of the Linnean Society* 120: 274–285 (2017)

Tsai, C.-C., R.A. Childers, N. Nan Shi, C. Ren, J.N. Pelaez, G.D. Bernard, N.E. Pierce, and N. Yu. "Physical and behavioral adaptations to prevent overheating of the living wings of butterflies." *Nature Communications* 11: 551 (2020)

ORGANIZATIONS DEDICATED TO THE STUDY AND CONSERVATION OF BUTTERFLIES

Butterfly Conservation (UK)
butterfly-conservation.org

Conservation of Butterflies in South Africa
cbisa.co.za

European Butterflies Group
butterfly-conservation.org/in-your-area/european-butterflies-group

Lepidopterists Society of Africa
lepsocafrica.org

McGuire Center for Lepidoptera and Biodiversity
floridamuseum.ufl.edu/mcguire

Monarch Watch (USA)
monarchwatch.org

Moths and Butterflies Australasia
maba.org.au

North American Butterfly Association
naba.org

The Lepidopterists Society (USA)
lepsoc.org

Xerces Society for Invertebrate Conservation (USA)
xerces.org

USEFUL WEBSITES

BugGuide
bugguide.net

Butterflies Australia
butterflies.org.au

Butterflies of Asia
butterfliesofasia.com

Butterflies of India
ifoundbutterflies.org

Butterflies of North America
butterfliesofamerica.com

Butterflies and Moths of North America
butterfliesandmoths.org

Butterflies and Moths of West Africa
westafricanlepidoptera.com

eButterfly
e-butterfly.org

iNaturalist
inaturalist.org

Learn about Butterflies
learnaboutbutterflies.com

Moths and Butterflies of Europe and North Africa
leps.it

UK Butterflies
ukbutterflies.co.uk

INDEX

ACKNOWLEDGMENTS

My life and the lives of butterflies have been entwined for a lifetime. I thank my parents, Alan and Doreen, who encouraged and steered me in my passion all those years ago. I thank my wife, Tanya and daughters, Jasmine, Rhiannon, and Annabella, for helping with butterfly studies and I am grateful to early mentors, especially Peter Payne and Michael Clark, who taught me how to be a naturalist and scientist. I also thank all the lepidopterists I have learnt from particularly Robert M. Pyle, David Nunnallee, Jon Pelham, and Lincoln Brower. I also thank Kate Shanahan and Kate Duffy and the rest of the UniPress team for ensuring the book in your hands is as lovely as it is.

David G. James

I'm grateful to my parents and all of the research advisors I've had over the years, including Kelly D.M. McConnaughay, May R. Berenbaum, Roger L. Kitching, Rudolf Meier, and especially my Ph.D. advisor, collaborator, and friend Naomi E. Pierce. Current and former members of my lab, including Kwaku Aduse-Poku, Jing V. Leong, Weijun Liang, Renato Nunes, and Susan Tsang (among others), keep me sane. A big thank you to my research collaborators, without whom my research would grind to a halt: Jade A.T. Badon, Perry A.C. Buenaventure, Michael F. Braby, Yu-Feng Hsu, Akito Y. Kawahara, Krushnamegh Kunte, Alma B. Mohagan, Alexander Monastyrskii, Tan Nhat Pham, Yi-Kai Tea, Djunijanti Peggie, Houshuai Wang, Zhengyang Wang, Shen-Horn Yen, and many others. Finally, many thanks to my partner, Bowen Zhang, for tolerating and sometimes encouraging my obsession with butterflies.

David J. Lohman

PICTURE CREDITS

The publisher would like to thank the following for permission to reproduce copyright material:

Alamy Photo Library: Aditya "Dicky" Singh 169; AGAMI Photo Agency 249; Andrew Waugh 120, 190–1; Arterra Picture Library 71; Avalon.red 114–15; Blickwinkel 11, 230B; C-images 15; D Guest Smith 162–3L; Dave Watts 93; DEEPU SG 22T; Dembinsky Photo Associates / Skip Moody 130; Dinodia Photos 159; Dominic Robinson 163R; Don Johnston_IH 36; Donna Ikenberry/Art Directors 99; Edward Parker 84; Fran Sanderson 265; Frank McClintock 229; George Grall 20; Ger Bosma 14; Gillian Pullinger 80–1L; Hideo Kurihara 128–129; Ian Redding 122; imageBROKER 149, 269; Jeff March 205; Larry Doherty 233B; Louise Heusinkveld 212; Minden Pictures 53; Nature Picture Library 91; Nigel Cattlin 234; Paul Weston 51; Pictorial Press Ltd 9; Ramachandran A 103B; Rasmus Holmboe Dahl 264; Rick & Nora Bowers 258; Sandra Standbridge 266; Skip Moody/Dembinsky Photo Associates 207; Survivalphotos 87; Tina Scott 273

Creative Commons: Gayatri dutta CC BY-SA 4.0 256; Sandipoutsider CC BY-SA 4.0 137; Scott O'Neill CC BY-2.5 238

Darlyne Murawski: 95

David Nunnalee: 126, 156

David G. James: 21, 22B, 24, 25, 27T, 28, 33, 34, 35, 37, 39, 40, 44, 46, 49, 55, 59, 63, 64, 65, 73, 74, 75, 78, 82, 83, 88, 89, 103, 105, 111, 113T, 125, 127, 131, 133, 134, 138, 139, 140, 142, 147, 153, 154–5, 158, 160, 161,165, 166, 167, 168, 171, 177, 179, 181, 183, 192, 193, 194, 195, 196, 210, 211, 213, 214, 215, 218, 219, 224, 226, 230T, 232, 233, 236, 237C, 240, 241, 243, 263

David J. Lohman: 33, 70, 116, 117

Dreamstime: Jam Shahzad 261; Matee Nuserm 38; Melinda Fawver 221; Pnwnature 220

Flickr: Andrew Allen 277; Dennis Holmes 53; Mike Friel 67; Rob Santry 259; rsa_nature 110; Tim Worfolk 237R

Gan Cheong Weei: 179

Ian Riddle: 251

iStock: adamkaz 255; AlbyDeTweede 175; Bob Hilscher 227; Court Whelan 62; CreativeNature_nl 245; DeeWard 181; espy3008 79; EstuaryPig 145; HeitiPaves 203; Ian_Redding 239; JillLang 228; kickers 66; marcouliana 19; MarekUsz 164; Martin Leber 27R; MikeLane 45, 121B ; MirekKijewski 19; Musat 19; randimal 188L; rbiedermann 19; Robbie Ross 136; SHAWSHANK 26, 61; Sundry Photography 107L; Tanja_G 121T; williamhc 112; yhelfman 196

Dr Yi-Kai Tea: 62, 151

Linda Kappen: 235

Nanfang Yu: 135

Nature Picture Library: Ingo Arndt 172L; Jussi Murtosaari 81R, 106, 108; Matthew Maran 109; Nick Garbutt 68; Piotr Naskrecki 102

Panos Pictures/Field Museum of Natural History, Chicago 254

Sciencephoto Library: B.G. Thomson 41

Shutterstock: Anant Kasetsinsombut 129; Andi WG 79; Angela N Perryman 124; alslutsky 19; Amalia Gruber 32; Anton Kozyrev 18; catocala 7, 19; chanwangrong 69; Cosmin Manci 19; Dave Montreuil 123; Davide Bonora 170C; Ekky Ilham 104; Erin Donalson 167; Francisco Herrera 216; Freebird7977 10; hisari45 197; Jason Benz Bennee 201; Jennifer Miner 85; Jim Nelson 262; Julee75 119R; Kendall Collett 199; Kirsanov Valeriy Vladimirovich 113B; Lara Bostock157; Leena Robinson 48; Lioneska 267; monster_code 231; Dr Morley Read 79; Nicole Patience 170L; Peter Waters 18; Radka Palenikova 188R; Rene Bechard 143; Ritam - Dmitrii Melgunov 23B; Rob D the Pastry Chef 247; Ronald Caswell 118; SanderMeertinsPhotography 271; Sandra Caldwell 97; Sandra Standbridge 257; Sari ONeal 8, 261B; Scott F Smith 77; Seema Swami 219T; Sergio Ortiz 12–13; Stephan Morris 275; Stephanie Kindermann 173; Super Prin 19; Tommy Daynjer 217; Tony Mills 57; yhelfman 24R;

Tanna Knouse 86

The Royal Society 45

Wikimedia Commons: böhringer friedrich 107R; Sharp Photography 23T, 251

William Piel and Antónia Monteiro 207

All reasonable efforts have been made to trace copyright holders and to obtain their permission for use of copyright material. The publisher apologizes for any errors or omissions in the list above and will gratefully incorporate any corrections in future reprints if notified.